00-ECO-5

WHOLE EFFLUENT TOXICITY TESTING: STUDY PLAN FOR FIELD VALIDATION

by:

Jerome M. Diamond, Ph.D.
Corinne Marino
James Stribling, Ph.D.
Tetra Tech, Inc.

2003

Water Environment Research Foundation

The Water Environment Research Foundation, a not-for profit organization, funds and manages water quality research for its subscribers through a diverse public-private partnership between municipal utilities, corporations, academia, industry, and the federal government. WERF subscribers include municipal and regional water and wastewater utilities, industrial corporations, environmental engineering firms, and others that share a commitment to cost-effective water quality solutions. WERF is dedicated to advancing science and technology addressing water quality issues as they impact water resources, the atmosphere, the lands, and quality of life.

For more information, contact:
Water Environment Research Foundation
635 Slaters Lane
Alexandria, VA 22314
Tel: (703) 684-2470
Fax: (703) 299-0742
www.werf.org
werf@werf.org

This report was published by the following organization. For non-subscriber sales information, contact:

IWA Publishing
Alliance House, 12 Caxton Street
London SW1H 0QS, United Kingdom
Tel: +44 (0) 20 7654 5500
Fax: +44 (0) 20 7654 5555
www.iwapublishing.com
publications@iwap.co.uk

Library of Congress Catalog Card Number: 2003104090

WERF ISBN: 978-1-893664-74-6 IWAP ISBN: 978-1-84339-667-3

Tetra Tech, Inc.

The research on which this report is based was funded, in part, by the United States Environmental Protection agency through Cooperative Agreement No. CR827345-01 with the Water Environment Research Foundation (WERF). Unless an EPA logo appears on the cover, this report is a publication of WERF, not EPA. Funds awarded under the Cooperative Agreement cited above were not used for editorial services, reproduction, printing, or distribution.

ACKNOWLEDGMENTS

Several scientists provided technical input in designing this study. We thank the Methods and Data Comparability Board, National Water Quality Monitoring Council, USEPA's Biological Advisory Committee, Florida Department of Environmental Protection, Hampton Roads Sanitation District, Joseph Gully, Deborah Denton (USEPA Region 9), Chuck Hawkins, Tom Cuffney, Trevor Reynoldson, Steve Canton, and Ron Klauda with the North American Benthological Society, and Jeroen Gerritsen of Tetra Tech for their valuable comments. We also wish to acknowledge Brenda Decker of Tetra Tech, Inc., who was responsible for producing this report.

Report Preparation

Principal Investigators:
Jerome M. Diamond, Ph.D.
James Stribling, Ph.D.
Corinne Marino, B.S.
Tetra Tech, Inc.

Project Team:
Jeroen Gerritsen, Ph.D.
Tetra Tech, Inc.

Project Subcommittee

Robin L. Autenrieth, Ph.D.
Department of Civil Engineering, Texas A&M University

Rodney W. Cruze
City of Riverside, California

Thomas W. La Point, Ph.D.
University of North Texas, Institute of Applied Science

Michael A. Lewis, Ph.D.
Office of Research & Development, U.S. Environmental Protection Agency

William Stubblefield
ENSR Corporation

Water Environment Research Foundation Staff

Project Manager: Dean E. Carpenter, D.Env.

BENEFITS

- The study plan summarizes available performance characteristics and data quality attributes for standard EPA acute and chronic freshwater WET tests and stream bioassessment methods.
- The study plan identifies appropriate methods and/or quality control criteria to compare laboratory and field results, using a data quality objectives approach.
- Several national or international organizations were involved in the development of the study plan, furthering ties with WERF and technical environmental issues of importance to wastewater dischargers.
- The study plan, and the results of the study when implemented, may provide regulators and dischargers valuable information with which to design more appropriate WET permit monitoring requirements.
- When implemented, the proposed research will produce additional data on methods performance for bioassessments and WET that may be applied to other environmental monitoring and data management efforts (e.g., state monitoring under Clean Water Act section 305(b), National Environmental Methods Index or NEMI).

Key Words: data quality objectives, measurement quality objectives, biological assessment, effluent-dependent streams, method performance

ABSTRACT

Under the National Pollution Discharge Elimination System (NPDES), many municipal and industrial wastewater treatment facilities must perform Whole Effluent Toxicity (WET) Testing. Regulatory agencies determine the level of compliance of each facility by making inferences from the results of these tests. There has been some concern regarding appropriate ways to integrate WET tests into NPDES permits, including how to relate WET tests and instream biological conditions. Previous research (Diamond et al., 1999) has examined this issue using historical data. Because of issues with data comparability, i.e. questionable data quality, and project design, results were inconclusive. This study plan was designed to collect new data on method performance for both WET and bioassessment that would help answer the outstanding question. The study plan was designed using a Data Quality Objective (DQO) approach in which DQOs and measurement quality objectives (MQOs) were defined. These DQOs and MQOs were developed using technical input from many scientists from federal, state, and private organizations. The DQO development process highlighted technical design issues that needed further investigation before implementing a definitive study. Among these issues were determining if DQOs and MQOs were achievable, and determining appropriate biological assessment methods for various ecoregions (e.g., effluent-dependent streams in the arid U.S. west). In order to appropriately address these issues, it was determined that a pilot study would be implemented before the full-scale study. The pilot study is designed as a one-year effort in which participating facilities will perform quarterly WET tests (using *Ceriodaphnia*, *P. promelas*, *Selenastrum*) and at least one bioassessment (including macroinvertebrate, fish, and algal assemblages) as well as providing other prescribed data. Results of the pilot will provide answers to technique design issues and will ultimately determine the most appropriate study design for the definitive study.

TABLE OF CONTENTS

LIST OF TABLES

LIST OF FIGURES

LIST OF ACRONYMS

BOD	Biochemical oxygen demand
BRS	Biological response signatures
CI	Confidence interval
C.V.	Coefficient of variability
DE	Discrimination efficiency
DQO	Data quality objective
IBI	Index of biological integrity
IC_{25}	Concentration of effluent that results in a 25% inhibition
IEC	Instream effluent concentration
IWC	Instream waste concentration
LC_{50}	Lethal concentration to 50% of the test organisms on average
LOEC	Lowest observed effect concentration
MQO	Measurement quality objective
MSD	Minimum significant difference
NABS	North American Benthological Society
NOEC	No observed effect concentration
NPDES	National Pollution Discharge Elimination System
PE	Performance evaluation
POTW	Publicly owned treatment works
PTD	Percent taxonomic disagreement
QA	Quality assurance
QAPP	Quality assurance project plan

QC	Quality control
RBP	Rapid bioassessment protocol
RMSE	Root mean square error
RPD	Relative percent difference
SD	Standard deviation
SI	Stressor identification
TSS	Total suspended solids
USEPA	United States Environmental Protection Agency
WET	Whole effluent toxicity
WWTP	Waste water treatment plant

EXECUTIVE SUMMARY

Introduction

Since implementation of the federal Clean Water Act, measures have been taken to protect and restore water quality and to ensure that designated beneficial uses of surface waters are met. The Clean Water Act specifies that point sources such as effluent discharges must not have adverse effects on beneficial uses, where adverse effects includes impacts on the biological condition of receiving streams where aquatic life uses are indicated. Whole Effluent Toxicity (WET) tests have been required by regulatory agencies as part of the National Pollution Discharge Elimination System (NPDES) permits for many facilities, to make inferences about instream biological conditions: if a WET test fails certain compliance criteria, then receiving water may be impaired or in danger of being impaired.

While WET tests have been regarded by many as a useful monitoring tool, there continue to be concerns as to appropriate ways to integrate WET test results into NPDES permits. A central issue in this debate is whether WET compliance limits are effective predictors of an effluents' influence on instream biota. Previous retrospective research examining this issue (Diamond et al., 1999) provided inconclusive results due to a variety of factors including: insufficient data, biased site selection, and uncertain data quality. One recommendation from that research was that new data be collected using current WET and bioassessment methods and quality assurance protocols to specifically evaluate this issue. The present research addresses this need through a consensus-based study design to test relationships between freshwater WET test results and biological condition in wadeable streams.

Project Objectives

The overall goal of this study was to evaluate the relationship between WET test results and instream biological conditions for receiving waters. Stated as a hypothesis: when WET tests indicate that there should be instream effluent toxicity, biological impairment is observed within a reasonable time. If WET tests indicate that there should be no toxicity instream, biological impairment is not observed.

Three specific project objectives were defined:

- Identify effluent discharge conditions and WET endpoints for which WET results are predictive and/or related to field biological results.
- Identify underlying disparities in WET test and instream results if and when they arise.
- Identify and characterize uncertainties associated with WET testing and biological assessments.

Project Scope

Although the overall project goal and objectives could encompass a variety of aquatic systems, types of WET tests, and wastewater dischargers, to make the study feasible, the project scope was limited to include the following criteria:

- Study sites are limited to wadeable, freshwater streams (not large rivers), as bioassessment methods are most standardized for those types of systems. In addition, a large and more comprehensive database is available for freshwater WET tests and biological assessment methods with which to interpret results of the study.
- Toxicity tests to be used are current USEPA effluent "chronic" toxicity tests for freshwater samples. These include 7-day tests using the daphnid *Ceriodaphnia dubia* and fathead minnow *Pimephales promelas,* and the four-day *Selenastrum* algal growth test.
- Because of limited resources, preference is given to those sites that currently perform toxicity tests and/or bioassessments.

Data Quality Objectives

A challenge was to design a workable plan, i.e., one that is cost-effective yet produces sufficiently rigorous data to satisfy study objectives. To address this challenge, the research team used EPA's data quality objectives (DQO) approach. Data quality objectives are quantitative and qualitative statements that describe the purpose of the study, specify the most appropriate data to collect, and determine the most appropriate methods and conditions under which to collect the data. In addition to the WERF Project subcommittee, recommendations on DQOs and other parts of the study design were solicited and provided by many state, federal, and private organizations including: the National Methods and Data Comparability Board, EPA's Biological Advisory Committee, National Water Monitoring Council, and key scientists affiliated with the North American Benthological Society.

The following DQOs were characterized for WET testing and bioassessment:

Accuracy. Method-dependent tests such as WET and biological assessment yield data that depend on how the method is performed. Therefore, accuracy cannot be independently determined for these methods. However, relative accuracy can be obtained for WET using standard performance (or blind reference toxicant) samples for which results have been demonstrated by a number of labs using the same method. WET accuracy will be demonstrated by having WET endpoints within +/- 30% of the mean endpoint value (i.e., LC_{50} or IC_{25}) for performance evaluation (PE) samples. Type I and Type II error rates will be assessed as part of the pilot study using a variety of measurement quality objectives (MQOs) (see below).

For biological assessment methods, accuracy is measured in terms of whether a given assemblage is assessed correctly; e.g., impaired, marginally impaired, not impaired. Accuracy will depend on adequate data or measurement quality throughout the different parts of the assessment protocol. Overall biological assessment accuracy will be demonstrated by having discrimination (classification) efficiencies of ≥85%, using sites with known stressors.

Power. Power is the ability of a given test or analysis to distinguish an effect at some prescribed level with a certain level of confidence. For WET, the DQO for power is that tests are able to statistically detect a 25% decrease in organism response (i.e., LOEC = IC_{25}) 90% of the time with 90% confidence.

For biological assessments, the DQO for power is that a biological assessment is able to detect ≥15% change in composition metrics and composite index values 90% of the time with 90% confidence.

Representativeness. The degree to which WET and biological data collected are representative of a given facility is critical to the outcome of this project. Previous research (Diamond et al., 1999) recognized this as one of the major limitations of earlier attempts to relate WET and instream biotic integrity. For WET testing, this means sufficiently frequent testing and/or other information regarding WET variability over time. The DQO for WET is that results represent overall toxicity of an effluent during the course of a year.

For biological assessments, representativeness is ensured by selecting assemblages and methods that are able to demonstrate the types and proportions of biota the stream supports.

Measurement Quality Objectives

Within any measurement system, there are inherent sources of variability. WET testing and bioassessment have the potential for systematic and random error that can cause misinterpretations of results. For purposes of this study, MQOs were identified to help control error sources in sampling and analysis and to ensure achievement of DQOs. MQOs are qualitative or quantitative criteria, containing specific units of measurement, that determine the quality of data needed in the study. Typically, these include measures of sensitivity, precision, and bias that serve to identify, isolate, and reduce the source of error, thus controlling data variability. Because bioassessments are not discrete methods but a series of methods grouped together, it is important to control the sources of variability with each step of the protocol. MQOs suggested for bioassessment and WET testing are shown in Tables ES-1 and ES-2, respectively.

Pilot Study Design

Because many of the specified DQOs have never been rigorously tested before, it was recommended that a pilot study be performed so that further research could address several technical issues, including:

- Are the proposed DQOs and MQOs achievable?
- How should biological data be interpreted for sites that are effluent-dependent and have no upstream (e.g., sites in the arid west)?
- What types of biological assemblages should be sampled to assess effluent effects?
- What is the minimum amount of WET and bioassessment data needed to adequately address study DQOs?
- What are the appropriate sampling methodologies for macroinvertebrates?

In order to meet the DQOs suggested above, the consensus process indicated that the following data are required:

- Biological measures for fish, macroinvertebrates, and periphyton up and downstream of an effluent discharge, or at every location sampled (where no upstream occurs).
- Appropriate reference condition information, if applicable
- Instream physiochemical measures, hydrological data, habitat quality assessment, effluent dilution at the time of WET testing
- Chronic WET tests: *Ceriodaphnia*, *Pimephales promelas*, and *Selenastrum capricornutum*
- Data for effluent permit parameters (e.g., chemical results required by permitting agency), effluent flow, wastewater hydraulic retention time, record of upsets/bypasses during the study
- Surrounding land use information

The pilot also was designed to help finalize the design and resource needs of a full-scale study, and to provide useful methods performance information for commonly used WET tests and biological assessment methods. Other data requirements or sampling designs may be identified for the full study, depending on the results obtained from the pilot.

Site Selection. For purposes of the pilot, certain site selection criteria were used to identify acceptable sites:

- No other point sources and preferably no obvious nonpoint sources, in the downstream sampling zone. This will help to ensure that any instream effects observed are due to the effluent and not another source.

- The effluent tested must be similar to what is discharged to the stream (i.e., no laboratory post-treatment such as dechlorination of the effluent prior to testing).

Table ES-1. Components of biological assessment protocols and approaches for controlling variability **in streams** (systematic and random error).

Assemblage Relevance							
Benthos	Fish	Periphyton	Potential Error Source and Type	Proposed Error Control Procedure	Performance Characteristic Addressed	DQO or MQO	References
Field sampling							
X	X	X	Patchiness of organism distribution (random error)	Large multihabitat sample	Representativeness	Standard protocol	Barbour et al., 1999
X			Degraded habitat/missing subhabitat (systematic error)	Consistent level of sampling effort/proportional distribution of sampling effort	Precision (intra-team and inter-team)	Metric value: ≤30% Index score: ≤15%	MDEQ, 2003
X	X		Unique habitat or habitat type: nonrepresentative of site class or region (systematic error)	Stratification	Precision	Within-class variance ≤ among-class variance	
X	X	X	Recency of storm event/high flows (systematic and random error)	Postpone sampling for 2-3 days, duration of postponement depends on channel size	Representativeness	Standard sampling decision criteria	
X	X		Over- or under-sampling of subhabitats (systematic error)	Sampling effort (benthos) allocated to subhabitats in proportion to frequency of occurrence in defined sampling reach; sampling (fish) covers all habitat in sampling reach	Accuracy	Statement: Sample produces data representing biota that overall stream habitat has the capacity to support	

Table ES-1 (continued). Components of biological assessment protocols and approaches for controlling variability **in streams** (systematic and random error).

Assemblage Relevance							
Benthos	Fish	Periphyton	Potential Error Source and Type	Proposed Error Control Procedure	Performance Characteristic Addressed	DQO or MQO	References
Field sampling (cont.)							
X	X	X	Inconsistency of sampling (inter-team; random error)	Training, habitat definitions	Precision	Metric value: ≤40% Index score: ≤20%	MDEQ, 2003
X	X	X	Inconsistency of sampling (intra-team, systematic error)	Training, habitat definitions	Precision	Metric value: ≤30% Index score: ≤10%	MDEQ, 2003
Sorting and subsampling							
X		X	Over- or under-representation of selected taxa (systematic error)	Fixed count/randomized subsampling, remove all organisms contained in aliquots	Accuracy, bias, representativeness	Standard subsampling protocol	Barbour et al., 1999; MDEQ, 2003
X			Non-representative taxa (systematic error)	Homogenization of sample, random selection of aliquots	Representativeness	Standard subsampling protocol	Barbour et al., 2003
X			Missed specimens (systematic error)	Sorter training/QC checks of sort residue	Bias (of sorter)	Sorting efficiency ≥90%	Barbour et al., 1999; MDEQ, 2003
X			Non-representative aliquot (systematic error)	Homogenization of sample; compare aliquot to total sample	Accuracy (of aliquot)	Relative percent difference (RPD) ≤5% (for composition metrics between aliquot and total sample)	MDEQ, 2003

Table ES-1 (continued). Components of biological assessment protocols and approaches for controlling variability **in streams** (systematic and random error).

Assemblage Relevance							
Benthos	Fish	Periphyton	Potential Error Source and Type	Proposed Error Control Procedure	Performance Characteristic Addressed	DQO or MQO	References
Taxonomy							
X	X	X	Mis-identification - taxonomic level (random error)	Designate more coarse (higher) taxonomic level as analytical requirement	Accuracy	Standard identification protocol	
X	X	X	Mis-identification - incorrect taxon (systematic error)	Use recognized taxonomy and technical literature	Accuracy	Standard verification protocol	MDEQ, 2003
X	X	X		Use highly trained and experienced taxonomist; or recognized research expert			
X	X	X		Identified and verified reference collection			
X		X	Mis-identification - incorrect taxon (systematic error)	Re-identification of randomly-selected samples by 2nd laboratory (benthos), or duplicate samples collected at individual locations and split to different taxonomists	Precision	Percent taxonomic disagreement (PTD) ≤ 12	MDEQ, 2003

Table ES-1 (continued). Components of biological assessment protocols and approaches for controlling variability **in streams** (systematic and random error).

Assemblage Relevance							
Benthos	Fish	Periphyton	Potential Error Source and Type	Proposed Error Control Procedure	Performance Characteristic Addressed	DQO or MQO	Reference
Enumeration							
X			Laboratory rough count grossly divergent from final count	Develop "rules" for what should be counted, and what should not be counted (rules should be slightly different for laboratory rough count vs. taxonomist final count): e. g., for invertebrates - *do not count* empty gastropod shells, damaged body parts without head and thorax attached, worm fragments other than cephalic end, embryonic asselids, etc.); if final count is >±20% of target number, re-pick (at random) from sort residue or unsorted sample remains; for final count, rely upon taxonomist, not sorter; for this cell, need input from periphyton specialist	Accuracy	Not applicable	
X		X	Specimen loss, over-clearing of slide-mounts	Improve attention to detail	Precision	RPD ≤5%	MDEQ, 2003

Table ES-1 (continued). Components of biological assessment protocols and approaches for controlling variability **in streams** (systematic and random error).

Assemblage Relevance							
Benthos	Fish	Periphyton	Potential Error Source and Type	Proposed Error Control Procedure	Performance Characteristic Addressed	DQO or MQO	Reference
Data entry							
X	X	X	Data entered incorrectly into database (systematic error)	Handchecking by comparing 100% of entered values to original datasheets	Accuracy	0% data entry errors	MDEQ, 2003
X	X	X	Incorrect data entered into database (systematic and random error)	Spot-checking of data recording routine	Accuracy	no intentional or inadvertent data recording errors	
Metric calculation							
X	X	X	Errors in computer-generated metric values/metric calculation algorithms (systematic error)	Randomly select 10% of calculated values, recalculate by hand; correct discovered errors and recalculate metrics	Accuracy	0% incorrect calculations	MDEQ, 2003
Site assessment and interpretation							
			Type I error: metric and index values indicate impaired biological condition when sites have not been impacted by stressors	Ensure MQOs are met for all phases of bioassessment	Precision or sensitivity	at 90% confidence, be able to detect ≤15% change	
			Inability of biological index to correctly detect chemically, physically, or hydrologically-stressed sites	Use only indices that have been properly calibrated	Accuracy	discrimination efficiency (DE)≥85%	MDEQ, 2003

Table ES-2. Proposed Measurement Quality Objectives (MQO) for WET

Potential Source of Variability	MQO	References
Intra-Test Precision	MSD for *P. promelas* chronic growth test and Selenastrum growth test < 35% MSD for *C. dubia* chronic test reproduction < 37%	Grothe et al.,1996; USEPA, 2000; ASTM 2000
Inter-Test Variability	Use laboratories with documented QA programs, successful track record Use consistent dilution series for facility C.V. for monthly reference toxicant test endpoints < 35% Duplicate WET tests quarterly: C.V. for endpoints < 35%	Grothe et al.,1996; USEPA. 2000; Warren-Hicks et al., 2000; Moore et al., 2000.
False Positives/ False Negatives	Split effluent testing with another participating lab: < 35% C.V. in IC25. Bi-monthly method blank tests: no significant effect compared to control ($p < 0.05$) Positive control elicits a significant response compared to control ($p < 0.05$) Associated reference toxicant test yields endpoint within 35% of the long-term average.	USEPA, 2000; Moore et al., 2000

The pilot study will involve participation from at least six wastewater treatment facilities that have met the prescribed selection criteria: three sites in the arid west region of the U.S. and three in the eastern temperate region. Additional or different facilities may participate in the pilot once implemented. The financial obligations of the pilot will be met through a collaborative process, involving the Water Environmental Research Foundation, each participating facility, and perhaps other funding sources. Each facility will provide at least the resources necessary to implement the testing and/or biological assessments they normally perform.

WET and Biological Methods

The duration of the pilot study will be one year. Because one of the objectives of the pilot is to determine which assemblages (or combination of assemblages) are most appropriate for assessing effluent effects, WET tests and biological assessment will examine three assemblages: invertebrates, fish, and algae. Using these three assemblages should help ensure that data collected for a given site are truly representative of the biota that the stream supports. Each bioassessment will be performed at least once during the year at an appropriate time period (index period) to maximize sampling gear efficiency and biological representativeness. Bioassessments will use EPA's revised Bioassessment Procotols. A habitat assessment will be performed concurrently with the biological assessment to help differentiate water quality from habitat effects.

Each facility will perform quarterly chronic WET testing using *Ceriodaphnia*, *Pimephales promelas*, and *Selenastrum*. The most current EPA protocols and quality assurance methods will be required. In addition, several quality control procedures will also be required to help ensure high (or at least known) quality data. These include: testing occasional blind blanks (non-toxic samples), duplicate tests, split sample effluent testing with another participating laboratory, and positive controls within each test.

Sampling Design

Stream sampling will be performed within three km above the discharge and within three km below the edge of the chronic mixing zone or the dilution point dictated by the compliance limit for WET testing. Many of the sites in the arid west region are effluent-dependent streams where there is no upstream. The pilot study will attempt to address this issue using several methodologies, including:

- Taking sequential downstream samples and evaluating if there is any change with distance from the discharge site (i.e., improvement in biotic integrity)
- Compiling available macroinvertebrate and fish data from other arid west sites (provided by the Arid West Water Quality Research Project, states, EPA regions, other WERF projects, municipalities, universities, and consultants).
- Evaluating the feasibility of developing stressor-tolerance ratings for widely distributed species in the arid west.

Pilot Study Evaluation

In order to address pilot objectives, an evaluation process will be implemented that will involve analyzing:

DQOs and MQOs. The appropriateness of DQOs and MQOs will be evaluated by comparing the actual bioassessment and WET performance with data quality acceptance criteria for each facility. Performance will be analyzed with respect to precision, accuracy, and QC parameters to determine how much usable data was obtained from each facility meeting DQOs and MQOs in the study design.

Redundancy among Bioassessment Types. In order to determine which biological assemblages should be sampled to assess effects, the degree of redundancy across assessment types (i.e., which types of assessments tend to yield similar results) will be analyzed. This analysis will involve comparing different biological assessment results for each individual facility and across facilities, as well the degree of variation in bioassessment results with facility toxicity and chemistry.

Comparison of Macroinvertebrate Sampling Methodologies. In order to determine which sampling methodologies are desirable for the full study, the pilot study will compare results obtained from two macroinvertebrate assessment techniques (standard grab sampling and artificial substrates. This comparison will include analyzing the precision and discrimination efficiency of each method. The results of both assessment methods will be compared with WET results for each facility to determine whether one is more highly correlated with WET than the other. The field performance of both methods will be compared to determine if, and under what circumstances, future sites in the full study would benefit from using one or the other method.

Preliminary Design for the Full Study

Pending results of the pilot study, DQOs and MQOs will be re-evaluated and revised if necessary. Although the types of data collected in the full study are likely to be similar to those collected in the pilot, certain bioassessment or WET methods may not be used pending their ability to meet the necessary DQOs.

Chief differences between the pilot and the full-scale study are that the full study will have:

- more facilities/sites with a range of treatment types and regional representation
- a wider range and greater representation of different effluent dilution categories
- a wider range of stream sizes (orders), excluding large rivers
- similar habitat quality in upstream and downstream sampling zones, if applicable
- a broader range of WET histories for facilities (e.g., some with no apparent toxicity historically, some with sublethal toxicity only).

CHAPTER 1.0

INTRODUCTION

It has been generally assumed that whole effluent toxicity (WET) tests provide some indication of potential instream effects, so long as exposure measures are taken into account (Grothe et al., 1996; LaPoint and Waller, 2000); that is, if WET tests indicate toxicity at actual effluent concentrations instream, impairment of instream biological conditions is likely. The principal confounding factors in attempts to perform comparisons (e.g., EPA's Complex Effluent Testing Program in the early 1980's) have been insufficient data, biased site selection, and questions on data quality, for both WET tests and instream biological assessments (Sprague, 1995; Marcus and McDonald, 1992; Waller et al., 1996; Diamond and Daley, 2000). Within the regulated community, there is a profound need to determine if WET and field biological integrity are related (Diamond and Daley, 2000; Grothe et al., 1996). Furthermore, data are needed to determine whether or not results obtained in the laboratory, using standardized WET tests, can be extrapolated to the field with confidence (Cairns, 1986).

The type and rigor of quality control activities necessary for WET testing and biological assessment can be identified by evaluating each assessment process as a series of discrete activities (Diamond et al., 1996; Barbour et al., 1999; Stribling et al., in review). By examining the individual steps, variability and potential error sources can be partitioned and minimized. The approach for performing WET testing is relatively well defined (e.g., USEPA, 1995; ASTM, 2000); i.e., there are prescribed operational procedures, test organisms, sample holding times, exposure conditions, and statistical analysis routines. However, there are a number of flexible steps in WET testing (e.g., food in *Ceriodaphnia* tests). Thus, variability of WET test procedures can be partitioned such that many, but not necessarily all, error sources are managed (Warren-Hicks and Parkhurst, 1996; Warren-Hicks et al., 2000). The protocols for performing field biological assessments, whether for periphyton, benthic macroinvertebrates, or fish, also consist of a series of methods, many of which can vary among organizations (e.g., states) and even within an organization (e.g., the Environmental Monitoring and Assessment Program, EMAP, versus the Mid-Atlantic Highlands Assessment, both within USEPA). To help control potential systematic and random error in WET and bioassessment data, and to determine the level of certainty that can be associated with each type of data, strict methods definition and standard operating procedures are developed for this study.

The proposed study will address data comparability and quality through the use of data quality objectives (DQO) and measurement quality objectives (MQO) for (a) collection and analysis of WET and instream biological assessment data, and (b) comparison of WET and biological assessment results. The goal of this exercise is to ensure that the final comparison between the two results is not hindered by questions regarding excessive variability or bias in the data. Appropriate quality assurance and quality control (QA/QC) can help reduce and manage method variability, although such variability can never be eliminated fully. However, *only* data that meet quantitative and qualitative MQO/DQO requirements for collection and analysis will be used in the WET/instream comparisons.

1.1 Technical Input

Scientists from different disciplines and organizations provided technical input on DQOs and MQOs for the Study Design. These organizations include:

- Methods and Data Comparability Board
- National Water Monitoring Council
- EPA Biological Advisory Committee
- Florida Department of Environmental Protection
- Los Angeles Sanitation District
- Hampton Roads Sanitation District
- USEPA Region 9
- A field focus group coordinated within the North American Benthological Society (NABS)
- Scientists from other state, private, and federal organizations.

Collaboration with the Methods and Data Comparability Board. The field focus group within NABS had a workshop in 2001 with the National Methods and Data Comparability Board (the Methods Board) to discuss the study design and provide technical input on field aspects of the study design. It was with their assistance that MQOs were proposed. At the NABS 2002 conferences held in May 2002, a special session was created on performance-based methods. As one of the members of the science advisory panel, the Methods Board provided technical expertise on characterizing method performance for WET and bioassessment. Both the Water Environment Research Foundation (WERF) and the Methods Board recognize the need for field and laboratory assessment protocols for biological and toxicological methods. It was partly out of this need that the Methods Board has concentrated its efforts on assessing data comparability and developing guidance for method performance. The Methods Board has recently developed a web-based database of environmental monitoring methods called the National Environmental Monitoring Index (NEMI). NEMI contains peer-reviewed chemical, biological, and radiochemical method summaries for regulatory and non-regulatory water quality analyses. Critical data are provided on the sensitivity, accuracy, precision, technique, and relative cost for each method. The outcome of this study is of particular interest to the Methods Board as it has the potential to provide useful information on method-performance that could be incorporated into NEMI and other programs.

1.2 Pilot Study

Research performed in this project, and by other workers, indicates that MQOs may not be well characterized for biological assessments and other field assessment protocols. The same is true, to a lesser extent, with WET test data. Therefore, this research recognized that a pilot study needs to be performed prior to a full study. The pilot study will help answer some outstanding study design issues, including:

- Determine whether the DQOs and MQOs proposed for the full study are achievable and appropriate
- Determine appropriate biological sampling methodologies for different regions (e.g., arid west vs. eastern temperate streams)
- Determine the types of biological assemblages that should be sampled to assess effluent effects
- Determine the minimum amount of data needed to address study objectives.

The pilot will provide important information concerning performance of commonly used WET tests and bioassessment procedures, and will provide a useful product for a wide range of environmental scientists and managers. In addition, the pilot study will be an invaluable tool with which to design the full-scale study and attract more facilities to participate.

The purpose of this document is to summarize the results of the DQO/MQO process for determining standard procedures and the types of data quality needed. This document also summarizes the proposed pilot study as well as the full study design.

1.3 Data Quality Objectives Approach

Before collecting environmental data, it is important to first determine the type, quantity, and quality of data needed to meet the project objectives and support a decision based on the results of data collection and observation. Study plans should be carefully prepared to best meet the project objectives (Figure 1-1) to avoid the risk of expending too much, not enough, or the wrong type of effort on data collection.

The DQO process developed by EPA (GLNPO, 1994; USEPA, 2000a) is an adaptive planning tool that systematically addresses these issues. The purpose of the DQO process is to improve the effectiveness, efficiency, and defensibility of decisions made based on the data collected, and to do so in a productive manner (USEPA, 2000a). This process specifically describes the uses of the data (e.g., the type of decision that needs to be made based on data collected) as well as other factors that will influence the type and amount of data to be collected (e.g., the problem being addressed, existing information and available resources).

In general terms, DQOs are qualitative and quantitative statements that clarify the purpose of a monitoring study, define the most appropriate types of data to collect, and determine the most appropriate methods and conditions under which to collect them. Additionally, DQOs define requirements for accuracy, precision, bias, method sensitivity, and range of conditions over which a method yields satisfactory data (Klemm et al., 1990). The products of the DQO process are: (1) acceptance criteria for data quality and (2) a data collection design that ensures that data will meet the criteria specified.

DQO's are developed in 3 stages: (1) determine what information is needed and why and how that information will be used; (2) determine methodological and practical constraints and technical specifications to achieve the information desired; and (3) compare different available methods and choose the one that best meets the desired specifications within identified practical and technical limitations (USEPA 1984, 1986, 1995; Klemm et al., 1990).

The use of specific agreed-upon DQOs at the beginning of the study design process will ensure several benefits to the outcome of this research. First, the DQO process is a consensus process that incorporates active stakeholder desires and expert information. Thus, before the study begins, organizations or agencies that have a large stake in the outcome of the research (in this case USEPA, states, and wastewater dischargers as represented by the Water Environment Federation and WERF) have agreed to the ways in which data are to be collected, analyzed, and interpreted.

A second major benefit of the DQO process is that it forces all parties to specify, up front, the kind and quality of data that need to be collected *to answer the stated questions*. Inherent to this process are stated MQOs or method performance characteristics that need to be met during the study. Performance characteristics include accuracy, sensitivity, precision, bias, and the range of conditions (temporal and spatial) over which the method is expected to yield appropriate data.

Step 1.	**State the Problem** Define the problem, identify the planning team, examine budget, schedule.
Step 2.	**Identify the Decision** State decisions, identify study questions, define alternative actions.
Step 3.	**Identify the Inputs to the Decision** Identify information needed for the decision (information sources, basis for Action Level, sampling/analysis methods).
Step 4.	**Define the Boundaries of the Study** Specify sample characteristics, define spatial/temporal limits, units of decision making.
Step 5.	**Develop a Decision Rule** Define statistical parameter (mean, median), specify Action Level; develop logic for action.
Step 6.	**Specify Tolerable Limits on Decision Errors** Set acceptable limits for decision errors relative to consequences (health effects, costs).
Step 7.	**Optimize the Design for Obtaining Data** Select resourceeffective sampling and analysis plan that meets the performance criteria.

Figure 1-1. The data quality objective process (USEPA 2000a)

CHAPTER 2.0

IDENTIFY THE DECISION TO BE MADE (STUDY OBJECTIVE): DATA QUALITY OBJECTIVES AND STUDY BOUNDARIES

The overall goal of the proposed research is to answer the following question:

"Do results obtained from laboratory WET testing agree with those from field-based instream biological assessments?"

Stated as a hypothesis: When WET tests indicate that there should be instream effluent toxicity, biological impairment is observed within a reasonable time. If WET tests indicate that there should be no toxicity instream, biological impairment is not observed.

2.1 Project Objectives

There are three specific project objectives associated with the overall study goal stated above:

- identify effluent discharge conditions and WET endpoints for which WET results are predictive of /or related to field biological results
- identify underlying causes for disparities in WET and instream results if and when they arise
- identify and characterize uncertainties associated with both WET testing and biological assessments.

2.2 Project Scope

The scope of the overall project goal is broad and could encompass many kinds of aquatic systems, types of WET tests, and a variety of wastewater dischargers. To make the project more feasible, in terms of both financial and human resources, the research team has limited the project scope as follows:

- Study sites will be limited to wadeable, freshwater streams (not large rivers) only, as bioassessment methods are most standardized for these types of systems and some data may already exist for several sites.
- Toxicity tests to be used are current USEPA effluent toxicity tests for freshwater effluent samples. These include chronic tests using the planktonic invertebrate *Ceriodaphnia dubia,* the fish *Pimephales promelas* (fathead minnow) and the 4-day *Selenastrum* algal growth test.
- Resources for performing this study are assumed to be limited. Therefore, preference would be given to those sites that are performing toxicity tests and/or bioassessments. Also, it is understood that all U.S. regions of the country or all habitat types cannot be represented in the study design, although attempts will be made to include as many as possible in the full study. The pilot study will examine two major types of discharge situations: arid western and temperate eastern U.S.

2.3 Data Quality Objectives (DQOs)

The following DQOs represent the types and quality of data recommended for the proposed pilot study and, pending the results of that study, the full study. The quantitative statements in these DQOs are based on published information obtained through this research and the experience of the research team performing these analyses.

Accuracy: Method-dependent tests such as WET and biological assessments yield data that depend on how the method is performed. Therefore, absolute accuracy cannot be determined for these methods. However, relative accuracy can be obtained for WET using standard performance (or blind reference toxicant) samples for which results have been demonstrated by a number of laboratories using the same method. Type I and Type II error rates have been estimated for some WET methods but it is not clear that these are appropriate as DQOs (see Section 6.2). Therefore, WET accuracy will be demonstrated by having WET endpoints within ± 30% of the mean endpoint value (i.e., LC_{50} or IC_{25}) for performance evaluation (PE) samples (USEPA, 2000b). EPA's recent WET interlaboratory study suggests that this accuracy criterion can be met by experienced laboratories for these tests (USEPA, 2000c).

Because accuracy is defined here as nearness to a "true" value (Keith 1991), our true value is the number of sites identified as stressed sites on the basis of chemical and physical water quality data. For biological assessment methods, accuracy is measured in terms of the frequency with which the overall index or individual metric correctly identifies a site as impaired. For example, if there is a dataset of 100 sites with known, documented stressors (i.e., 100 = the true value), and there is indication of biological impairment at 75 of those sites, the accuracy of the index is 75%. Accuracy of an index will depend on adequate data or measurement quality throughout the different parts of the assessment protocol. Based on several studies performed by Tetra Tech for state programs (Wyoming: Stribling et al., 2000; Maryland: Snyder et al., 1998; and Mississippi: MDEQ, 2002), overall biological assessment accuracy will be demonstrated by having discrimination (classification) efficiencies of ≥ 85%, using sites with known stressors.

Power: WET tests are able to detect a 25% decrease in organism response 90% of the time with 90% confidence.

Biological assessments are able to detect ≥15% change in composition metrics and composite index values 90% of the time with 90% confidence.

Representativeness: WET results represent overall toxicity of an effluent during the course of a year.

Field sampling methods provide data on the overall biota that the stream supports.

CHAPTER 3.0

IDENTIFY INPUTS TO THE DECISION (DATA REQUIREMENTS)

Several types of data will be needed in the pilot study design to accomplish the objectives of this research and address the data quality objectives identified:

- Biological measures (fish, macroinvertebrates, and periphyton) up and downstream of an effluent discharge, or at every location sampled (where no upstream exists)
- Appropriate (i.e., calibrated) reference condition information, if applicable
- Instream physicochemical measures, hydrological data, habitat quality, effluent dilution/flow at the time of WET testing
- Chronic WET tests: *Ceriodaphnia* (invertebrate), *P. promelas* (fish) and *Selenastrum* (alga)
- Data for effluent permit parameters measured, flow, hydraulic retention time, record of upsets/bypasses during the study
- Surrounding land use information

Pending the results of the pilot study, other parameters may be defined in the full study design.

The following chapter briefly describes major sources of variability associated with the two data components of this research, bioassessment and WET, and the ways in which these sources of variability will be addressed. Chapter 5 discusses proposed MQOs to ensure procedures for reducing variability identified in Chapter 4, yield data quality appropriate for the DQOs established. Chapter 6 summarizes the proposed design for the pilot study, which incorporates procedures to address systematic and random sources of error for WET and bioassessment identified in Chapter 4. Finally, Chapter 7 discusses the current proposed design for the full study, pending results of the pilot study.

CHAPTER 4.0

SOURCES OF VARIABILITY AND PROCEDURAL SOLUTIONS

There are generally two types of error, systematic and random, that cause variance in the results of any measurement system (Taylor, 1988) (Figure 4-1). However, because there is often interaction between these two error sources, they should not be examined as if they were completely independent of each other. For ecological studies, random error often results from natural variability for which controls are either unspecified or difficult to construct. Random error in the current study would include WET variability over time and changes in effluent dilution over time. Many sources of random error, however, can be managed through design parameters, as discussed in later sections of this report. For both WET testing and field biological assessments, systematic or measurement error is controlled through development and use of a system of quality control (QC) activities that serve to identify, isolate, and reduce or eliminate potential error sources in the sampling and analysis process, thus controlling data variability.

4.1 Field Biological Assessments

While there are multiple indicators of ecological health used in bioassessments, benthic macroinvertebrates are the most commonly used assemblage for stream ecosystems, and there has been extensive research and development of these methods (Southerland and Stribling, 1995; Gibson et al., 1996; Birge et al., 1989; Eagleson et al., 1990; Dickson et al., 1992; Barbour et al., 1999). The Intergovernmental Task Force on Monitoring Water Quality (ITFM, 1995) recommended using at least two assemblages in assessing the quality of aquatic habitat because one assemblage is less likely to be sensitive to all effluents and contaminants. Furthermore, one assemblage type may not reflect all possible impairments, particularly for those effluents that have multiple contaminants with different modes of action. Targeting multiple species over different trophic levels is more likely to improve detection capability over a broader range of stressors than a single assemblage because each trophic level composing the aquatic community would be expected to have a distinct response range to perturbation events or degraded conditions (Barbour et al., 1995).

However, sampling and assessment methods for some types of flora or fauna (e.g., fish) may be challenging, making it difficult to meet the types of DQO and MQOs proposed for this research. For example, the mean generation time and mobility of individual species may affect the degree to which WET results can be related to instream biological results: species with shorter generation times (e.g., most algal species) generally have higher population turnover and recruitment rates, which tend to mask effects of a short-term (e.g., days-weeks) toxicity episode. Similarly, those species that are mobile (e.g., most fish) could avoid short-term toxicity episodes and recolonize quickly when water quality conditions improve. In both cases, the assessment may not capture actual biological effects instream. Given the current uncertainties concerning which biological assemblages are most appropriate, the three main biological assemblages monitored by federal and state agencies will be used in the pilot study at each site: fish, macroinvertebrates, and periphyton (attached algae). Pending the amount of unique information obtained from each assemblage type, the full study may or may not evaluate all three assemblages.

Biological assessment protocols are not methods, but rather are a series of methods grouped together (see Figure 4-2). Thus, the performance characteristics associated with each component contribute to the overall uncertainty associated with the final assessment (Diamond et al., 1996; Stribling et al. [in review]). Therefore, potential error sources need to be controlled for each step of the protocol. Table 4-1 lists the components of biological assessment protocols and approaches for controlling variability (random and systematic error).

4.2 WET Testing

Several sources of systematic error or method variability for WET tests have been identified and are summarized in Table 4-2. Factors such as test organism sensitivity (which may be a function of culturing techniques and diet), type and chemical quality of dilution water, and analyst experience/technique are major sources of variability in WET testing (Warren-Hicks et al., 2000; Moore et al., 2000; USEPA, 2000b). These factors may vary widely depending on the specific EPA protocol and which version of the protocol a given laboratory is using (Markle et al., 2000). Variability in short-term chronic toxicity tests is further influenced by the choice of endpoint and the statistical analysis used to derive the endpoint (Baird et al., 1996; Bailer et al., 2000). The interpretation of statistical tests, the validity of assumptions, and the choice of ways in which each laboratory reports WET, will be addressed in the study. The following is a list of general steps proposed to address WET variability.

- Use the most recent USEPA methods protocols for chronic tests.
- Require standard dilution water - either reconstituted or dilute mineral water - hardness and pH should be similar to that in the receiving stream.
- Between and within analyst variability - personnel that conduct tests must be the same personnel that routinely conduct the WET tests at that laboratory facility; personnel must be identified clearly and consistently in records.
- Within-test variability - include a positive control and/or reference toxicant test side-by-side with the WET test. Require both a minimum and a maximum C.V. and/or MSD for each test type (see Chapter 5).
- Within-lab variability - perform duplicate tests on the same effluent sample.
- Between-lab variability - test occasional split samples among participating laboratories.
- Effluent Variability Effects - test multiple effluent facilities representing a range of ecoregions or physiographic provinces, hydrologic regimes, dilution, and habitat

quality (Diamond and Daley, 2000); test facilities with different levels of chemical variability (e.g., different retention times or equalization capacities).

Software used in statistical analysis will be reported and a comparison of results using different statistical software will be performed during testing. Two or three nationally recognized software packages that are used to analyze WET data will be recommended (e.g., Toxcalc, Tidepool Software, Inc.).

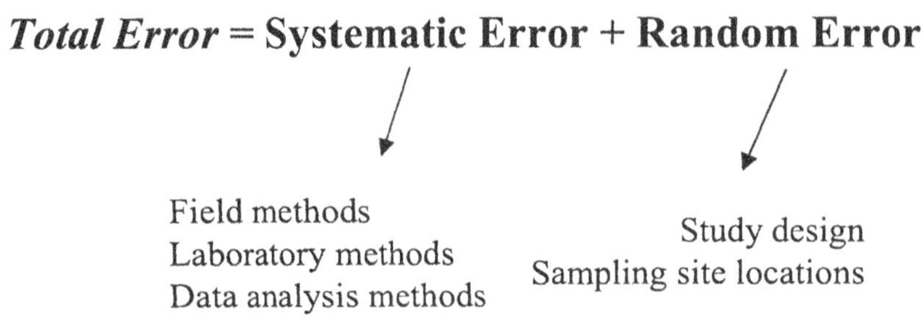

Figure 4-1. Components of decision error or variability in environmental data, and aspects of study components that effect them.

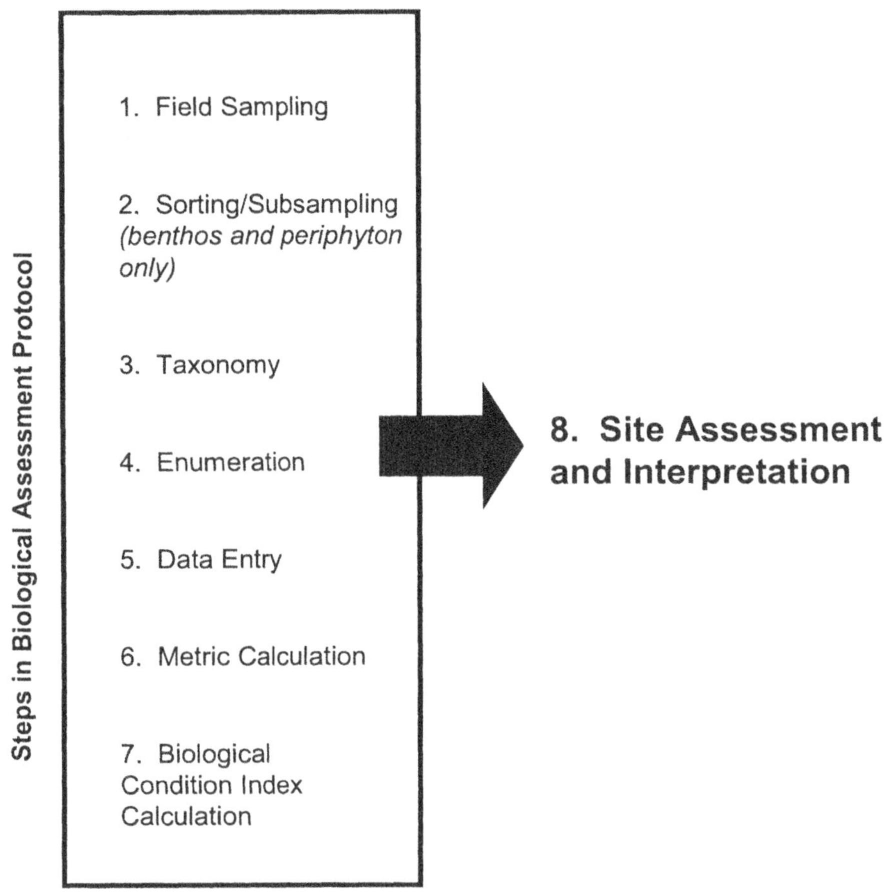

Figure 4-2. Components or steps of the biological assessment process.

Table 4-1. Components of biological assessment protocols and approaches for controlling variability in **streams** (systematic and random error).

Assemblage Relevance							
Benthos	Fish	Periphyton	Potential Error Source and Type	Proposed Error Control Procedure	Performance Characteristic Addressed	DQO or MQO	References
Field sampling							
X	X	X	Patchiness of organism distribution (random error)	Large multihabitat sample	Representativeness	Standard protocol	Barbour et al., 1999
X			Degraded habitat/missing subhabitat (systematic error)	Consistent level of sampling effort/proportional distribution of sampling effort	Precision (intra-team and inter-team)	Metric value: ≤30% Index score: ≤15%	MDEQ, 2003
X	X		Unique habitat or habitat type: nonrepresentative of site class or region (systematic error)	Stratification	Precision	Within-class variance ≤ among-class variance	
X	X	X	Recency of storm event/high flows (systematic and random error)	Postpone sampling for 2-3 days, duration of postponement depends on channel size	Representativeness	Standard sampling decision criteria	
X	X		Over- or under-sampling of subhabitats (systematic error)	Sampling effort (benthos) allocated to subhabitats in proportion to frequency of occurrence in defined sampling reach; sampling (fish) covers all habitat in sampling reach	Accuracy	Statement: Sample produces data representing biota that overall stream habitat has the capacity to support	

Table 4-1 (continued). Components of biological assessment protocols and approaches for controlling variability **in streams** (systematic and random error).

Assemblage Relevance							
Benthos	Fish	Periphyton	Potential Error Source and Type	Proposed Error Control Procedure	Performance Characteristic Addressed	DQO or MQO	References
Field sampling (cont.)							
X	X	X	Inconsistency of sampling (inter-team; random error)	Training, habitat definitions	Precision	Metric value: ≤40% Index score: ≤20%	MDEQ, 2003
X	X	X	Inconsistency of sampling (intra-team, systematic error)	Training, habitat definitions	Precision	Metric value: ≤30% Index score: ≤10%	MDEQ, 2003
Sorting and subsampling							
X		X	Over- or under-representation of selected taxa (systematic error)	Fixed count/randomized subsampling, remove all organisms contained in aliquots	Accuracy, bias, representativeness	Standard subsampling protocol	Barbour et al., 1999; MDEQ, 2003
X			Non-representative taxa (systematic error)	Homogenization of sample, random selection of aliquots	Representativeness	Standard subsampling protocol	Barbour et al., 1999
X			Missed specimens (systematic error)	Sorter training/QC checks of sort residue	Bias (of sorter)	Sorting efficiency ≥90%	Barbour et al., 1999; MDEQ, 2003
X			Non-representative aliquot (systematic error)	Homogenization of sample; compare aliquot to total sample	Accuracy (of aliquot)	Relative percent difference (RPD) ≤5% (for composition metrics between aliquot and total sample)	MDEQ, 2003

Table 4-1 (continued). Components of biological assessment protocols and approaches for controlling variability **in streams** (systematic and random error).

Assemblage Relevance							
Benthos	Fish	Periphyton	Potential Error Source and Type	Proposed Error Control Procedure	Performance Characteristic Addressed	DQO or MQO	References
Taxonomy							
X	X	X	Mis-identification - taxonomic level (random error)	Designate more coarse (higher) taxonomic level as analytical requirement	Accuracy	Standard identification protocol	
X	X	X	Mis-identification - incorrect taxon (systematic error)	Use recognized taxonomy and technical literature	Accuracy	Standard verification protocol	MDEQ, 2003
X	X	X		Use highly trained and experienced taxonomist; or recognized research expert			
X	X	X		Identified and verified reference collection			
X		X	Mis-identification - incorrect taxon (systematic error)	Re-identification of randomly-selected samples by 2nd laboratory (benthos), or duplicate samples collected at individual locations and split to different taxonomists	Precision	Percent taxonomic disagreement (PTD) ≤ 12	MDEQ, 2003

Table 4-1 (continued). Components of biological assessment protocols and approaches for controlling variability **in streams** (systematic and random error).

Assemblage Relevance							
Benthos	Fish	Periphyton	Potential Error Source and Type	Proposed Error Control Procedure	Performance Characteristic Addressed	DQO or MQO	References
Enumeration							
X			Laboratory rough count grossly divergent from final count	Develop "rules" for what should be counted, and what should not be counted (rules should be slightly different for laboratory rough count vs. taxonomist final count): e. g., for invertebrates - *do not count* empty gastropod shells, damaged body parts without head and thorax attached, worm fragments other than cephalic end, embryonic asselids, etc.); if final count is >±20% of target number, re-pick (at random) from sort residue or unsorted sample remains; for final count, rely upon taxonomist, not sorter; for this cell, need input from periphyton specialist	Accuracy	Not applicable	
X		X	Specimen loss, over-clearing of slide-mounts	Improve attention to detail	Precision	RPD ≤5%	MDEQ, 2003

Table 4-1 (continued). Components of biological assessment protocols and approaches for controlling variability **in streams** (systematic and random error).

Assemblage Relevance							
Benthos	Fish	Periphyton	Potential Error Source and Type	Proposed Error Control Procedure	Performance Characteristic Addressed	DQO or MQO	References
Data entry							
X	X	X	Data entered incorrectly into database (systematic error)	Handchecking by comparing 100% of entered values to original datasheets	Accuracy	0% data entry errors	MDEQ, 2003
X	X	X	Incorrect data entered into database (systematic and random error)	Spot-checking of data recording routine	Accuracy	no intentional or inadvertent data recording errors	
Metric calculation							
X	X	X	Errors in computer-generated metric values/metric calculation algorithms (systematic error)	Randomly select 10% of calculated values, recalculate by hand; correct discovered errors and recalculate metrics	Accuracy	0% incorrect calculations	MDEQ, 2003
Site assessment and interpretation							
			Type I error: metric and index values indicate impaired biological condition when no stressors are present	Ensure MQOs are met for all phases of bioassessment	Precision or sensitivity	at 90% confidence, be able to detect ≤15% change	
			Inability of biological index to correctly detect chemically, physically, or hydrologically-stressed sites	Use only indices that have been properly calibrated	Accuracy	discrimination efficiency (DE)≥85%	MDEQ, 2003

Table 4-2. Potential Sources of Variability for WET Testing

Potential Source of Variability	Possible Ways to Reduce Variability	References
Within Test Factors	use in-house test organisms where possible; ensure proper handling techniques to minimize damage/stress to organisms; include a positive control and/or a reftox side-by-side; use of MSD criteria, require maximum C.V.	Grothe et al., 1996; USEPA, 2000 b; ASTM, 2000
Culturing Techniques and Diet	maintain healthy in-house cultures using as brood stock only individuals that exhibit ideal reproductive capacities (i.e., able to produce young prolifically); daily maintenance of cultures (feeding/changing of culture water); feed a carefully controlled recommended diet using food that meets USEPA guidelines; keep daily records regarding cultures	Grothe et al., 1996; USEPA, 2000b; ASTM, 2000
Culture/Testing Water	use USEPA guidelines on water preparation; test water after preparation (alkalinity, hardness, pH, conductivity) to ensure proper water quality; maintain proper distilled water conditions (filters, hoses, connectors, etc. all in proper working order) to optimize quality; maintain any holding tanks used in proper condition (i.e., cleaned and properly aerated); and periodically test culture/testing water for contaminants	USEPA, 2000b; ASTM, 2000
Implementation of Test Methods/Procedures	use experienced personnel and properly train any new personnel; double-check water quality measurements (e.g., conductivity, pH.) by a second trained technician to ensure proper conditions	USEPA 2000b; Moore et al., 2000
Variations Due to U.S.EPA Protocol Changes and Optional Aspects of Protocols	ensure that most recent, up to date USEPA protocol is being used for testing; limit optional steps for test methods	Markle et al., 2000.
Data Recording Activities	QA measures such as double-checking of entries by two different trained technicians; keeping accurate charts to record performance over time	USEPA, 2000b; ASTM, 2000

Table 4-2 (continued). Potential Sources of Variability for WET Testing

Potential Source of Variability	Possible Ways to Reduce Variability	References
WET Method Factors	reject results from laboratories that have unacceptably high variability in reftox results over time, a C.V. of between 30 and 40% should be the cutoff; make use of a comparison of entire concentration-response plots rather than rely on a C.V. for a given point estimate; incorporate additional national test acceptability criteria (e.g., upper and lower bounds of MSD); monthly method-blank WET tests used to calculate normal range of variation expected under non-toxic conditions; semi-annual duplicate tests	Warren-Hicks et al., 2000; Moore et al., 2000; USEPA, 2000
Inter-Lab Variability	Split testing of one effluent sample with another certified laboratory; split testing with a method blank (i.e., a known non-toxic sample), duplicate testing	Moore et al., 2000.
Study Design Factors		
Site Variability and Extent of Fluctuation in Chemicals	include a range of wastewater facilities with different retention times, equalization potential, influent consistency	Diamond et al., 1999
Representative Effluent Samples for WET Tests for a Facility	conduct frequent WET tests, using samples representative of effluent quality	Diamond and Daley, 2000
Effluent-Dilution, Receiving Stream Habitat and Ecology	need a range of effluent situations represented (low or high dilution, habitat types, hydrology, or climatic region)	Diamond et al., 1999; Barbour et al., 1996

CHAPTER 5.0

MEASUREMENT QUALITY OBJECTIVES

5.1 Biological Assessments

Table 4-1 shows categories of activities for which QC procedures and measurement quality criteria (or MQO's) will be established for biological assessment activities. Collection and analysis of duplicate samples from each sampling location will help quantify method and assessment precision (Stribling et al., 2001 [in review]). Previous work by Tetra Tech, Karr and Chu (1999), and several states has identified reasonable precision estimates for "replicate" stream samples and split duplicate samples. These precision values also will constitute proposed MQOs for the pilot study, because they have been shown to be achievable. The precision of metrics c scores for replicates will be evaluated to determine whether field and laboratory data quality can meet the MQOs proposed.

To determine the precision of taxonomic identifications, benthic invertebrate and algal samples will be identified by separate taxonomists or laboratories. Taxonomic precision will be determined by sending random, processed, identified samples from one laboratory to another expert for verification. The MQO for this precision is stated as the percent taxonomic disagreement (PTD).

To address data representativeness (i.e., data collected for a given site being truly representative of the full assemblage that the stream has the ability to support), all biological and algal samples (i.e., periphyton, macroinvertebrates, and fish) will be collected using a multi-habitat technique to ensure that all biota within an assemblage are sampled and evaluated. Sampling methods will follow USEPA's revised Rapid Bioassessment Protocols (RBPs) for wadeable streams (Barbour et al., 1999). However, two different macroinvertebrate sampling procedures (Hester-Dendy, an artificial substrate technique, and dip net sampling) have been used extensively across the U.S. (see Appendix B). The pilot study will compare results from macroinvertebrate assessments obtained by the two methods. To supplement biological data, habitat assessment data and instream physicochemical data (pH, dissolved oxygen, temperature, and conductivity) will be collected simultaneously, also adhering to Barbour et al., 1999.

5.2 WET Testing

As shown in Table 5-1, there are several MQOs that are addressed for WET testing in the study design. Key issues include test precision, false positives and false negatives (a statement of test accuracy), and sample representativeness. The following section summarizes considerations addressed within each MQO issue.

5.2.1 Measuring and documenting WET test precision

For chronic tests that use a hypothesis testing statistical design (e.g., *Ceriodaphnia* 7-day test), within-test variability can be measured by the minimum significant difference (MSD) value, which represents the amount of difference from the control that can be detected statistically using an analysis of variance or other hypothesis testing technique (USEPA, 2000b). MSD calculations may not be inappropriate for non-normal distributions and/or heterogeneous variances and are not relevant to point estimates (e.g., IC_{25} or LC_{50} values, see below and Grothe et al., 1996 and Warren-Hicks et al., 2000). The MSD criteria in this pilot design are based on a given cumulative percentile from a toxicity database using multiple laboratories (USEPA, 2000b; 2001). The use of MSD criteria should improve consistency between tests and between laboratories by limiting within-test variability. Tests not meeting MSD requirements would be qualified as suspect and repeated, if possible. Both a minimum and maximum MSD are stipulated for chronic tests to reduce Type I and Type II error rates (see below).

For point estimates (e.g., LC_{50} or IC_{25}), the 95% C.I. are used to document precision. This value is compared with the value derived from multiple tests conducted with the same test method and reference toxicant for each test and point (Grothe et al., 1996).

A further test of WET precision is to require duplicate WET tests at least once for each facility. Each replicate test will be a complete WET test, performed with different batches of the same effluent sample and organisms from the same batch.

5.2.2 Means of addressing false positives and negatives

In hypothesis testing as in chronic tests, balancing the potential occurrence of false positive/false negative results is a function of the number of samples collected, the variance of the sample mean (of the ecological endpoint), and the Type I and II error criteria selected by the investigator (Grothe et al., 1996). False positives are addressed in part by performing method blanks periodically (Moore et al., 2000). Therefore, this study recommends split testing of a method blank (a known non-toxic sample) with another certified laboratory semiannually. In addition, a semi-annual "PE" or reference toxicant solution will be sent to participating laboratories and treated as a normal effluent sample. Results for these samples (i.e., IC_{25}) should be ± 30% of the mean endpoint value for the toxicant. Split testing of effluent samples with another certified laboratory quarterly would also help document accuracy (Moore et al., 2000).

Type II errors in WET are largely controlled by documenting appropriate sensitivity and reliable test results through reference toxicant testing. For this study, the design calls for side-by-side reference toxicity testing with effluent tests and/or monthly reference toxicant tests and side-by-side positive controls, consisting of the average lowest observed effect concentration (LOEC concentration of a reference toxicant for the lab performing the tests.

In addition, Type I and Type II errors will be addressed in the pilot study through the use of upstream water (if available), or closely similar water but without effluent, as a second diluent in WET testing. At least once during the pilot study, each facility will perform side-by-side WET tests using standard lab water and stream water as diluents. Comparison of results, as well as the upstream water control (no effluent added) will be used to assess the degree to which lab water diluent provides representative WET results.

5.2.3 Sample Representativeness

Key factors in this study design pertaining to effluent sample representativeness relate to the frequency of testing (i.e., the probability of obtaining an accurate assessment of effluent quality over time) and the ability to capture the range of water quality types over time in the proportion in which they actually occur. Much of this discussion is presented in Chapter 7. However, there are several MQOs that specifically pertain to this issue as well.

First, in the pilot study, at least quarterly WET testing will be performed for each facility using all WET test types. Facilities that can do more frequent testing are preferred. For the full study, monthly acute and chronic WET tests are desired to capture effluent variability. Second, flow-proportional 24h composite samples are recommended in all tests and for sample renewals in chronic testing. Holding time is # 36h after sample collection for three of the days in chronic testing, including set-up on the first day. All samples must be received by the lab at # 4EC if shipped off-site (USEPA, 1995). Third, effluent testing will be randomly scheduled every quarter (or month) to further ensure sample representativenesss.

Table 5-1. Proposed Measurement Quality Objectives (MQO) for WET

Potential Source of Variability	MQO	References
Intra-Test Precision	MSD for *P. promelas* chronic test and *Selenastrum* growth test < 35% MSD for *C. dubia* chronic test reproduction < 37%	Grothe et al., 1996; USEPA, 2000; ASTM, 2000
Inter-Test Variability	Use laboratories with documented QA programs, successful track record Use consistent dilution series for facility C.V. for monthly reference toxicant test endpoints < 35% Duplicate WET tests quarterly: C.V. for endpoints < 35%	Grothe et al., 1996; USEPA, 2000; Warren-Hicks et al., 2000. Moore et al., 2000.
False Positives/ False Negatives	Split testing with another participating laboratory: < 35% C.V. in IC_{25} or LC_{50}. Split testing of method blank with another lab: C.V. in IC_{25} or LC_{50} < 35% Bi-monthly method blank tests: no significant effect compared to control ($p < 0.05$) Positive control elicits a significant response compared to control ($p < 0.05$) Associated reference toxicant test yields endpoint within 35% of the long-term average.	USEPA, 2000; Moore et al., 2000

CHAPTER 6.0

PILOT STUDY

As mentioned previously, a pilot study will be performed prior to a full study because there are several study design issues needing resolution (see below). First and formost, the pilot study results will determine if proposed DQOs and MQOs are feasible and whether alternate DQOs and/or MQOs need to be used in the full study. A side benefit of the pilot study is that it will yield current, rigorous information on how well freshwater WET tests perform in routine monitoring situations. Furthermore, the pilot study also may encourage other facilities to participate in the full-scale study and will enable WERF to finalize the full study design and its budgetary requirements.

6.1 Site Selection Criteria

The following criteria have been proposed for selection of acceptable sites (facilities) for the pilot study:

- Site must be freshwater, wadeable stream
- No point source or nonpoint source pollution 1 km downstream of the sampling zone
- Sites where the effluent tested is similar to what is discharged to the stream

Facilities will include those in four effluent dilution categories: < 25%, 25- 50%, 50 – 75% and > 75% effluent instream based on design low flow (i.e., 7Q10) conditions. As actual effluent dilution at the time of testing is required, facilities with access to accurate stream flow records (e.g., USGS gauge) are preferred. Facilities on streams with poor-to-fair instream habitat quality as well as better habitat will be included (Barbour et al., 1996a; Diamond et al., 1999). It is preferred to have facilities with a range of retention times and treatment processes so as to evaluate qualitative differences in effluent consistency over time.

The pilot study will examine sites only in the arid western and in the temperate eastern U.S; dischargers from additional regions may be incorporated into the full study, pending the resolution of certain design issues. Table 6-1 summarizes dischargers that have tentatively agreed to participate in a pilot study.

6.2 Pilot Study Design

Because of limited resources, the pilot has been limited to a one-year study. WET testing will involve chronic tests only, performed at least on a quarterly basis. Bioassessments will be performed at least once for each of the three assemblages: macroinvertebrates, fish, and periphyton. The timing of the bioassessment will be region specific. For the full study, it is desirable to conduct two assessments in each year (e.g., Spring and Autumn Index Periods) to better define data trends in data over time. Bioassessments performed during different seasons may ensure that different life history events are captured and potential population or community effects are more likely to be observed. Habitat assessment will be performed concurrently with biological assessment to differentiate chemical effects from habitat effects.

For the pilot study, sampling zones for dischargers with viable upstream flow will be limited to within three km above the discharge and within three km below the edge of the chronic mixing zone. This will ensure that downstream effects, if they occur, are properly identified with respect to either water quality effects of the discharge, habitat effects of the discharge, or other physiochemical effects independent of the discharge (see Figure 6-1 for sampling design). This design, however, presupposes a viable upstream for a facility. In many arid west (and some eastern U.S.) discharge situations, there is no upstream available for sampling. This issue will be addressed in several ways in this pilot (see below). A sampling design incorporating multiple stations (i.e., replicates) in each sampling zone increases the precision of the assessment, which, in turn, will increase the precision of the comparison between WET and biological results (Faith et al., 1991; Underwood, 1996; Barbour et al., 1996a; Diamond et al., 1999). Therefore, for the pilot study, where upstream sampling is feasible, two locations will be sampled upstream of the discharger and two locations in each of two downstream sites (Figure 6-1). Upstream locations will be positioned less than 0.5 km apart from one another, while two samples are to be positioned randomly within the chronic mixing (near complete-mix) zone. Having multiple upstream and downstream sites (replicates) will also help characterize natural variability when examining upstream-downstream comparisons (Smith et al., 1993; Bode and Novak, 1995).

For sites with upstream flow, acute and chronic mixing zone sites may be useful for examining biological condition as one moves farther away from the discharge and the concentration of effluent decreases. Mixing zone sampling will aid in quantifying the amount of biological impairment caused only by the effluent since the change in biological response can be compared to the change in effluent concentration. The use of multiple stations downstream also will be useful in assessing biological effects for sites with no upstream station. Furthermore, multiple downstream sampling will take into account differing state water quality regulations (i.e., presence or absence of mixing zones) on the study findings.

A desirable step in the bioassessment portion of the study design is to establish regional reference conditions (Barbour et al., 1999) based on aggregate measurements collected from several locations for each site class (groupings of sites that are similar in their innate biological potential). Reference sites should be chosen carefully to represent the least impacted sites within the region of study while possessing similar habitat features compared to the study area. Such reference sites are not easily identified or perhaps available for certain kinds of waterbodies, particularly effluent-dependent streams (such as in the arid west). These sites will require other approaches (see Section 6.3).

For those sites for which reference conditions are feasible (non-effluent dependent streams), it will be desirable to select sites in states or regions in which appropriately calibrated reference conditions have already been developed, because these can then be used to evaluate biological conditions of upstream, downstream, and mixing zone sites. States that have established reference conditions include Maine, New York, Ohio, North Carolina, Georgia, Mississippi, Oregon, Wyoming, Arizona, Florida, and Maryland. Again, for arid west sites, reference conditions are not easily determined. This concept will be addressed in the pilot study in other ways (see below). The use of both regional and upstream reference sites in the study will increase the certainty in quantifying the amount of impairment downstream caused by discharges alone.

6.3 Specific Issues to be Evaluated in the Pilot Study

6.3.1 Sites Without Upstream Flow

Some of the facilities participating in the pilot study will be located in the arid west region of the U.S. Many of the streams in this region are effluent-dependent, and may have no upstream. One of the important objectives of the pilot will be to evaluate the biology downstream of these facilities, using one or more of the following site-specific approaches:

- Take sequential downstream samples and evaluate if there is any change (i.e., improvement) in the biology as a function of distance from the facility, independent of habitat quality changes.
- Obtain available macroinvertebrate and fish data from other arid west sites, especially other effluent-dependent streams, and evaluate the use of simulated reference conditions for a given ecoregion of the arid west.
- Apply results from other studies; e.g., Tetra Tech is currently evaluating designated uses for effluent-dependent streams in California for USEPA Region 9 and the State. Tetra Tech will use other WERF and Arid West Water Quality Research Project (URS Corporation, CDM Inc., 2002) studies dealing with aquatic ecological issues in the arid west, as well as aquatic life use analyses to develop procedures to address reference condition for arid west discharge streams.
- Use measures related to the abundance of taxa either known or thought to be sensitive to water quality stressors. If the number or percent of stressor-sensitive organisms increases or decreases with distance downstream from a point source, and there are no substantial stressor inputs in the intervening river mileage, then the change in sensitive taxa likely is related to the point source. Clearly, this approach is feasible only if there are sufficient data to support tolerance values for many of the widely distributed species (algal, fish, and macroinvertebrate). Data exist in the arid west to build such a species tolerant database.

6.3.2 Use of Artificial Substrates

In general, the use of multi-habitat dip net sampling for benthos (Barbour et al., 1999) has been shown to be a comprehensive and easily adaptable sampling technique for most types of wadeable streams. However, some stream habitats (e.g., arid west sites) may not be amenable to sampling using standard dip net sampling. Sites could be difficult to sample due to shifting substrates (sand) and/or large boulders or bedrock. In these cases, artificial substrates may be a more appropriate sampling procedure. Table 6-2 summarizes advantages and disadvantages of using artificial substrates. For the pilot study, both dip net sampling and artificial substrates (Hester-Dendy: MDEQ, 2001) will be used at each site (see Appendix B for an explanation of the sampling protocols). This could be subject to change depending on the type of substrate and sampling conditions found at each site. Pending results of the two types of sampling on assessment results, the full study may or may not use both sampling procedures for macroinvertebrates.

6.3.3 Monitoring Periphyton Assemblages

Another pilot study objective is to determine the value added in incorporating algae in bioassessments. Monitoring periphyton, along with macroinvertebrates and fish, could give a more accurate picture of impairment status (R. Jan Stevenson, University of Michigan and S. Call, KYDEP, personal communicaton). Table 6-3 summarizes advantages of including periphyton, as well macroinvertebrate and fish groups for assessing water quality. Appendix B explains the sampling protocol for periphyton.

6.3.4 WET Testing

Determining the relationship between WET testing and the biological condition of receiving waters requires characterization of the frequency and magnitude of toxic events coming from the discharge. Toxic events are expected to be rare events, and routine monitoring may or may not capture them. Therefore, WET testing should be randomized so that estimates of the frequency of toxicity are unbiased. In addition to WET testing, it will be useful to have access to routine effluent monitoring information to determine the frequency of exceedances of conventional pollutants (BOD, TSS, chlorine), and the frequency of treatment upsets and related incidents.

Representativeness in terms of WET is dependent on sufficient testing over time (Diamond and Daley, 2000). For the pilot study, at least quarterly testing is required although more frequent testing is preferred. For the full study, it is hoped that facilities could perform monthly chronic WET tests (possibly including *Selenastrum*, depending on pilot results). Test dates for WET tests will be randomly assigned within each month or testing period for each discharger; acute endpoints for *C. dubia,* and *P. promelas* will be derived from the chronic tests using data collected by the lab at 48h.

A consistent dilution series should be used for a given type of test across time for a given discharger to make comparisons more feasible. For each test, a control and 6 effluent concentrations will be used. Five of these concentrations will reflect various dilutions of the permitted effluent instream waste concentration (see Table 6-4). Appendix A gives a brief overview of test condition and test criteria for each test organism (*Cerodaphnia, P. promelas, Selenastrum*).

While lab water (USEPA, 1995) will be the standard dilution water in WET testing, upstream water controls are also suggested to evaluate possible interactions between the effluent and actual stream water. If upstream water is not toxic to the test organism, it is suggested that stream water tests be performed side-by-side three times per year with lab dilution water tests.

6.4 Pilot Study Evaluation

As noted previously, a major objective of the pilot is to determine whether the proposed DQOs and MQOs are achievable and appropriate in a standard monitoring context. Therefore, the first part of the evaluation process will be to compare actual bioassessment and WET test performance for each facility with the data quality acceptance criteria proposed. This will include an analysis of:

- Precision: based on duplicate results and observed MSDs for WET by test type and RPDs, CVs, and RMSEs obtained for bioassessment by assemblage type (composition metrics and composite IBI)
- Accuracy: based on PE and split sample analyses testing in the case of WET and discrimination efficiencies obtained in bioassessments
- QC parameters: number and percent of invalid WET test results due to insufficient control performance; unacceptable WET test blank results (Type II error); sorting efficiency and taxonomic accuracy for biological assessments.

Analysis of each of the above parameters for WET and bioassessments will be performed to determine completeness of the data (i.e., how much usable data was obtained for each facility meeting DQOs and MQOs) and to identify possible modifications in DQOs or MQOs, or the study design, for consideration in the full study.

A second evaluation will compare the different biological assessment results for each facility and across facilities to determine the degree of redundancy across assessment types (i.e., which types of assessments tend to yield similar results). If it is demonstrated, for example, that periphyton bioassessments yield the same results for each site as the macroinvertebrate assessments, then the full study may incorporate the macroinvertebrate but not the periphyton bioassessment. A key to this evaluation will be not only the assessment results themselves, but also the degree of variation in results with facility toxicity and chemistry. An assessment method that varies little in results across sites is likely to be a relatively insensitive type of assessment and should not be used in the full study.

A third evaluation will compare the results of the two macroinvertebrate sampling techniques in terms of precision and discrimination efficiency. This information will be used to determine whether a sampling method should or should not be considered for use in the full study. Also, results of both assessment methods will be compared with WET results for each facility to determine whether one is more highly correlated with WET than another. Finally, the field performance of both methods will be compared to determine if and when future sites in the full study would benefit from using one or the other method.

6.5 Pilot Cost Estimation

The total estimated cost of the pilot study (Table 6-6) was calculated using unit prices for each analysis. Each unit price was estimated using the research team's past experience in performing these analyses, as well as input obtained from other laboratory facilities.

Most facilities tentatively identified already perform *Ceriodaphnia* and fish chronic testing (Table 6-1). Therefore, these facilities should continue to provide the cost of these analyses as part of the pilot study. Several arid west discharges also perform macroinvertebrate bioassessments, and those facilities likely will provide the cost of benthic bioassessments. Algal WET tests and periphyton and fish bioassessments are unlikely to be sponsored by participating facilities.

Most of the candidate pilot study facilities interviewed thus far have a design dilution of > 75%. In order to include a wider range of dilution categories in the study, an effort will be made to get participation from facilities with smaller dilutions. One source on information on additional appropriate facilities may be Dyer and Wang (2002), who have developed an extensive database of publicly owned treatment works (POTWs) in Ohio, including bioassessment, chemical, and spatial data.

For WERF to gain some insight into its financial responsibility to the project, Table 6-7 has been generated that lists the costs that are not likely to be provided by facilities.

6.6 Pilot Study QA and Project Management

Prior to initiating the pilot study, the Project Managers should develop a Quality Assurance Project Plan (QAPP) that discusses not only the data quality aspects contained in this report, but also names of participants, their contact and site information and all logistical procedures and steps to be followed in the Study. The QAPP will be developed in accordance with USEPA guidance (USEPA, 2001) and should be approved at a minimum by WERF (or its Project Sub-Committee), each participant in the study, and the Project Managers. Given the relationship of this study to wastewater regulatory requirements (i.e., WET testing), it would be desirable to have USEPA as a reviewer and signee of the QAPP as well. Once the QAPP is approved, the pilot study can be initiated. The Project Managers will oversee all aspects of the project to ensure that the study plan is followed by participants, that correct methods are used, and that all data are submitted for evaluation as scheduled. In addition, the Project Managers should coordinate all split sample WET testing and PE samples for WET analyses. The Project Managers should be responsible for all data analyses and the pilot evaluation, as well as reports to WERF. Finally, the Project Managers should ensure that all budgetary requirements are being met. The following subsections discuss more specific QA responsibilities that should be overseen by the Project Managers for bioassessment and WET testing.

6.6.1 Bioassessment

Project Managers should keep track of contact information for each facility and each sampling team. They should have records of all sampling locations, frequency of sampling, including a schedule of sampling for each participating facility. They should have a record of collection times, storage (preservation methods), and transport of samples. They should make sure that proper sample handling protocols are met for each facility; this should include sample log records and chain-of-custody forms. They also should have copies of all documentation for QC activities, including taxonomic bench sheets, subsampling rates, and sorting efficiency.

6.6.2 WET

Project managers should consult with laboratories performing the WET tests. They should have a record of the methods that each laboratory should use in WET testing including diluent used, effluent dilutions tested, and reference toxicant results. They should oversee QC requirements such as: inter-laboratory testing of blanks and samples, use of positive controls, and replicate results. Project Managers should check a percentage of the raw data for completeness, calculation errors, and data entry errors. All raw data should be entered into a master electronic database for analysis.

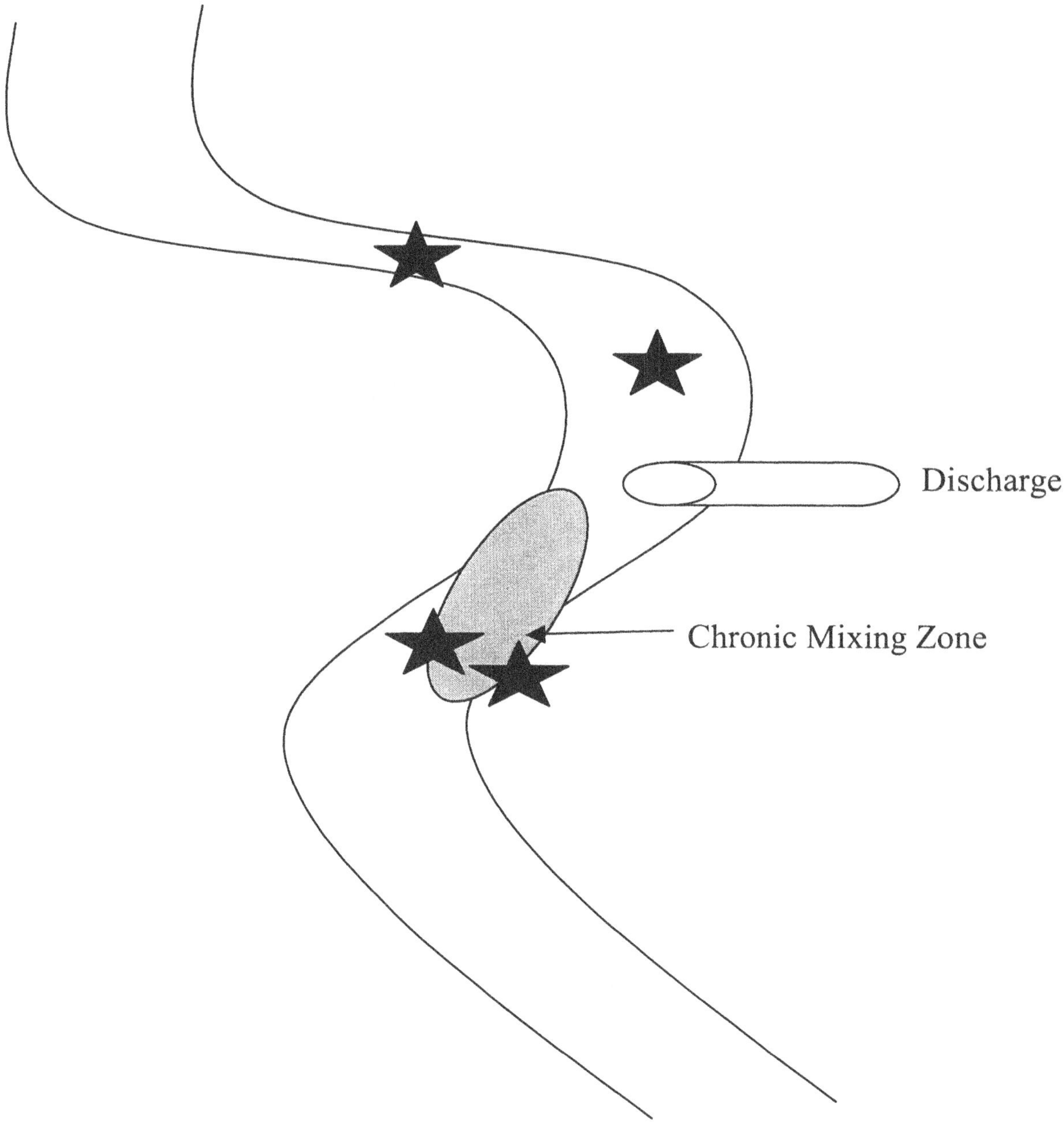

Figure 6-1. Example Sampling Design. (Stars indicate sampling locations.)

Table 6-1. List of Potential Dischargers for the Pilot Study

Contact	Discharger	State	Dilution Category	Testing Done at or for Facility	Comments
Kevin Oakes	**Goose Creek Taylor Run**	**Pennsylvania**	>75% >75%	**(1.)WET (*Ceriodaphnia*, fathead minnow)**	**Only Chronic WET**
Dan Koplish	**City of Allentown**	**Pennsylvania**	<25%	**(1.)WET (*Ceriodaphnia*, fathead minnow)**	**Only Chronic WET**
Todd Harris	**Denver Metro**	**Colorado**	>75%	**(1.)WET (*Ceriodaphnia*, fathead minnow) (2.)Bioassessment**	**Only Chronic WET**
Vicki Card	**Colorado Springs**	**Colorado**	>75%	**(50.) WET (*Ceriodaphnia*, fathead minnow) (2.)Bioassessment**	**Only Chronic WET**
Ed Curley/ Karen Sierra	**Pima County& City of Phoenix**	**Arizona**	>75%	**(1.)WET (*Ceriodaphnia*, fathead minnow, *Selenastrum*)**	**Only Chronic WET**
Dyer and Wang (2002)	**Ohio POTWs**	**Ohio**	**25-50% 50-75%**	**To be determined**	**Ohio EPA database and landuse info**

Table 6-2. Advantages/Disadvantages of Artificial Substrates (adapted from Barbour et al., 1999).

Advantages	Disadvantages
Artificial substrates allow sample collection in locationsthat are typically difficult to sample effectively.	Artificial substrates require a return trip; this may be a significant consideration in large states or those with limited technical resources.
As a passive sample collection device, artificial substrates permit standardized sampling by eliminating subjectivity in sample collection technique. Direct sampling of natural substrate requires similar effort and degree of efficiency for the collection of each sample.	Artificial substrates are prone to loss, natural damage or vandalism
Confounding effects of habitat differences are minimized by providing a standardized microhabitat.	Microhabitat standardization may promote selectivity for specific organisms if the artificial substrate provides a different microhabitat than naturally available at a site, i.e., the material of the substrate will influence thecomposition and structure of the community; solid artificial substrates will favor attached forms over motile forms and compromise the usefulness of the siltation index.
Sampling variability is decreased due to a reduction in microhabitat patchiness, improving the potential for spatial and temporal similarity among samples.	Orientation and length of exposure of the substrate will influence the composition and structure of the community.
Sampling collection using artificial substrates may require less skill and training than direct sampling of natural substrates.	

Table 6-3. Comparative Advantages of Different Taxonomic Groups for Use in Assessing Water Quality (adapted from Mackie, 2001).

	Algae	Macroinvertebrates	Fish
Distribution- value for assessing local conditions in lakes and streams	Omnipresent, but most useful in lentic habitats. Only attached forms occur in streams. Not useful in headwater areas.	Limited migration patterns. Hence, good indicators of localized conditions; well suited for site-specific impacts in both lake and streams. More diverse than fish in headwater streams.	Extensive migration abilities. More useful for assessing ecosystem impacts, except in headwater streams where fish fauna may be limited.
Response time- value for assessing short- and long-term effects.	Very rapid reproduction rates and life cycles. Most useful for short-term impacts.	Most species have 1-2 yr life cycle; integrate of effects of short-term environmental variations. Good for long-term analysis since historical databases are common and some (e.g. unionid clams) live >20 yrs for good long-term biomonitors.	Relatively long-lived and mobile. Excellent integrations of long-term (several years) effects.
Taxonomy- ease with which organisms can be identified, stability of taxonomy.	Requires a specialist in most cases, but often non-taxonomic methods are used (e.g. chlorophyll a, biomass)	Relatively easy to identify to family and genus. Heavy emphases on insects for which taxonomy is relatively stable.	Easy to identify to species. Taxonomy relatively stable for most families.
Sampling ease- ease with which communities can be sampled and resultant impact on resident organisms	Easy to sample, inexpensive and requires few people.	Relatively easy to sample. Requires few people, inexpensive gear. Recolonization by species removed is rapid, with little impact on resident biota.	Relatively easy to collect. If identified in field, fish can be returned unharmed.
Indicator value- Extent of documentation for taxa within each group.	Many species useful for assessing organic pollution especially for sewage. Directly affected by physical and chemical factors.	Extensive documentation on indicator value for nearly all phyla.	Salmonids, cottids, some percids especially useful for lake and stream assessments. Many cyprinids also useful in headwater streams.
Trophic position- Importance to humans	Autotrophs at bottom of food chain, vital to ecosystem functioning	Primary food source for fish of sport and commercial value.	Top of food chain. Most important group because human food.

Table 6-4. Proposed Study Design for WET Tests in the Pilot Study

Conduct testing for a 1-year period for each facility
Conduct at least quarterly chronic WET tests (*Ceriodaphnia, Selenastrum*, and *Pimephales*).
Use two dilution waters (upstream and lab) for testing side-by-side at least once during the year in order to compare results
Concentration series in chronic tests will consist of a control and 6 effluent concentrations which will reflect the permitted instream waste concentration (IWC) of effluents by bracketing the IWC in the following manner: (1) 100% effluent, (2) [IWC + 100]/2, (3) IWC, (4) IWC/2, (5) IWC/4, (6) IWC concentration at actual flow conditions at the time of sampling. Concentration series for each test type will remain constant for each discharger for entire testing period.
Test conditions will comply with the most recent USEPA protocol for chronic effluent testing
Test results suggesting noncompliance with permitted conditions would be repeated immediately
Facilities also will provide information on size and capacity of POTW, permit compliance effluent quality data; % flow rate of influent and effluent; residence time of effluent; pretreatment information

Data Analysis

Use the IC_{25} as a chronic endpoint, as well as calculating alternate IC values (e.g., IC_{20}, or IC_{30}) and NOEC; follow USEPA protocol for analysis of each endpoint.

QA/QC

Monthly chronic reference toxicant tests, and/or side-by-side reference toxicant tests with the discharger's test
Blank analysis at least once in the study
Use of a positive control (e.g., NaCl or copper concentration approximating the average LOEC) during testing unless a reference toxicant test is being done side-by-side
Split WET testing at least twice with another certified laboratory to determine accuracy of testing procedures and to assess WET variability, C.V. for *C.dubia* and *P.promelas* chronic tests should be #25% overall
Laboratories performing WET tests must have established QA programs and provide control charts reflecting acceptable reftox results; prefer labs that participated in USEPA's Interlaboratory Study, and/or State Certification to ensure that experienced personnel are involved in conducting these tests

Table 6-5. Proposed Study Design for Bioassessment in the Pilot Study

Study Design

Conduct at least one bioassessment for each facility
Sample fish, macroinvertebrate, periphyton assemblages
Sampling protocols must be comparable to Rapid Bioassessment Protocol methods, and have performance characterized
Each assemblage is sampled upstream and downstream of effluent discharge or at every location sampled (when no upstream occurs)
A habitat assessment will be conducted at each sample location
Facilities will also provide information on surrounding land use, physiochemical measures (if possible), and appropriate reference condition information (if applicable)

Data Analysis

Results will be analyzed by the ability to detect ≥ 15% change in composition metrics and composite values 90% of the time with 90% confidence

QA/QC

Duplicate taxonomy will be performed for 10% of samples for taxonomic precision
Duplicate field samples will be analyzed for field precision (10% of sites)
Duplicate subsamples will be taken from 10% of samples for subsampling precision/representation
QC checks will be performed on samples subsorted. Sorting efficiency ≥ 90%
100% of data entry will be QC'd- handchecked against original datasheets.

Table 6-6. Pilot Study Cost Estimate

Pilot Study Design	Number of analysis per facility	Cost per facility
*WET: *Ceriodaphnia*	8[1]	$ 9,600
*WET: *P. promelas*	8[1]	9,600
*WET: algae	8[1]	12,000
standard bioassessments[2]	1	10,000
Hester Dendy bioassessment[3]	1	7,000
Subtotal		**48,200**
QA/Project Mgmt		7,000
Data analysis		5,000
Cost (1 site)		**60,200**
Cost (6 sites)		$361,200

Assumed Unit Prices	
Analysis	**Cost/per sample**
WET (*Ceriodaphnia*, fish)	$ 1200
WET algae	1500
standard bioassessment	10000
Hester Dendy bioassessment	7000

* WET test cost includes monthly reference toxicant testing

[1] assumes four quarterly tests plus blank test, upstream dilution water test, split lab test and duplicate test

[2] fish, macroinvertebrate, and periphyton

[3] macroinvertebrates only

Table 6-7. Cost of the Pilot Study to WERF

Participating Facilities	Test not funded by facility	Cost to WERF
Goose Greek (West Chester)	1. WET algae	$ 12,000
	2. Bioassessment- macroinvertebrates, fish, periphyton*	17,000
Taylor Run (West Chester)	1. WET algae	12,000
	2. Bioassessment- macroinvertebrates, fish, periphyton*	17,000
City of Allentown	1. WET algae	12,000
	2. Bioassessment- macroinvertebrates, fish, periphyton*	17,000
Denver Metro	1. WET algae	12,000
	2. Bioassessment- fish, periphyton*	14,000
Colorado Springs	1. WET algae	12,000
	2. Bioassessment- fish, periphyton*	14,000
Pima County & City of Phoenix	1. Bioassessment-macroinvertebrates, fish, periphyton*	17,000
Subtotal		**155,000**
Project Management		42,000
Data Analysis		30,000
Total		**$227,000**

*** Bioassessment includes standard bioassessment and Hester-Dendy bioassessment**

CHAPTER 7.0

FULL STUDY DESIGN

7.1 Site Selection Criteria

The success of the full study in demonstrating the relationship between WET and bioassessment results depends in part on the overall representativeness of the study sites. As pointed out earlier in this report, a shortcoming of several earlier attempts was biased or non-random selection of sites. While the full study also is likely to focus on some dischargers, the *a priori* selection criteria will not include the performance of those facilities or previous results. Sites with previous WET or bioassessment results should be preferred, but the actual results will n be unknown until the facility is already selected for the study. At a minimum, 30-40 dischargers should be included in the study to ensure site representativeness. The site selection criteria are as follows:

- Sites will be selected *a priori* from a variety of U.S. ecoregions (e.g., arid west, mid-west eastern temperate, and Pacific northwest temperate).
- Sites should represent a range of dilutions, stream orders (but not large rivers), and habitats.
- Each site should contain no other point source or nonpoint source discharges within the recommended sampling zone.
- The instream habitat should be similar between upstream and downstream locations, if applicable.
- Effluent tested should be representative of what is discharged.
- A wide range of effluent toxic histories is preferred (i.e., some with historic chronic problems, acute problems, or no problems).
- Appropriate reference site data is desirable.
- Study sites are limited to freshwater wadeable streams.
- Monthly WET testing is preferred.

7.2 Proposed Data Analysis for Full Study Design

There are two general phases of analysis associated with this study: the first phase is analysis of biological and WET data individually to determine whether: (a) there is biological impairment due to the effluent and (b) WET test results indicate instream toxicity and therefore biological impairment. The second phase of analysis to determine the strength of association between WET and bioassessment results and sources of error. The following sections summarize the type of data and analyses included in each of these phases.

7.2.1 Phase 1 Analysis – Instream Biological Conditions

Bioassessment data need to be evaluated in a rigorous fashion to obtain robust, reliable results. This should be accomplished by incorporating minimum significant difference (MSD) analyses into the statistical approach of the study design to better determine "true" differences in organism response above and below the discharge. For two sites to be considered different, the metric values, scores, or multivariate distances between the sites would need to exceed the stated MSD objective.

Upstream and downstream (or "reference" and downstream in the case of sites with no upstream sampling location) biological results will be compared with each other. Degradation of stream condition will be assessed by comparison of upstream-downstream metric values (and available indices, if appropriate) using an analysis scheme similar to that used by Diamond et al. (1999). Any difference in metric or index values that (1) indicates poorer condition downstream and (2) is in excess of the 95th percentile of differences among upstream (or "reference") paired segments, will be scored as net degradation. Better condition that is similarly in excess of upstream (or "reference") paired segments will be scored as net improvement. Differences not in excess of upstream (or "references") paired segments will be scored as no change. For those sites that have both an upstream and true reference conditions, comparisons will be made between upstream and downstream conditions with references to determine the degree of upstream impairment. If both upstream and downstream samples exhibit biological impairment, as compared to calibrated reference conditions, the Stressor Identification (SI) process (USEPA, 2000d) will be implemented to determine which stressors are the most likely causes of the observed impairments (see Figure 7-1).

Data will be expressed as relative difference values (Diamond and Daley, 2000) for each metric and index score measured, as well as the overall assessment for each assemblage (fish, macroinvertebrates, periphyton) based on both upstream site precision values and the reference condition for the same ecoregion or habitat type.

Data from stream sites in Florida where duplicate samples were taken (Barbour et al., 1996b) provide a basis for estimating standard deviations for macroinvertebrate metrics and indices calculated with such metrics. Estimated standard deviations of metrics and a modified Florida index (Stream Condition Index, or SCI, on a 100-point scale) from repeat samples are shown in Table 7-1. The estimated standard deviation of individual metrics is from 7-18% of the range of the metric; and the standard deviation of the additive index is less than 6% of its range. If these estimates are assumed to hold for all repeated or adjacent samples, then a downstream site is statistically different from two upstream sites when its total score is more than 10 points from the mean of the 2 upstream sites. For adjacent upstream-downstream sites, 10 points, or 10% of the total range, is a minimum detectable difference at 90% confidence. These results are consistent with other repeated sample tests, which all suggest that a minimum detectable difference of index scores is approximately 10% of the total range (usually a 100-point scale).

7.2.2 Phase 1 Analysis – WET Predictions

WET data will result in estimates of frequency and magnitude of toxicity by species and endpoint type. Results will be normalized by effluent dilution at the time of testing. For each site, there will be actual WET endpoints (and 95% confidence intervals if appropriate) for each of the four types of tests, as well as several alternate "assessments" for each WET test denoting whether the test results predicts instream toxicity or not. Similar to the analysis scheme used in Diamond et al. (1999), individual monthly acute WET tests will be categorized as predicting either an effect (impairment/toxicity instream) or no effect (no toxicity instream) based on several different endpoint "triggers" commonly used in regulatory permitting: $LC_{50} < 100\%$ effluent or the 1Q10 dilution instream, percent mortality in 100% effluent < controls or < 20% mortality in 100% effluent (t-test, $p \leq 0.01$). In addition, acute data will be categorized on the basis of effluent dilution in the acute mixing zone at the time of testing. Monthly chronic WET data also will be assessed according to the methods used by Diamond et al. (1999). Individual chronic tests will be categorized as either probable instream toxicity or no toxicity based on commonly used endpoint triggers: NOEC < instream effluent concentration or IEC (7Q10 and actual dilution at time of testing), LOEC < IEC (design and actual), and IC_{25} < IEC (design and actual). In addition, this design will examine bioequivalency (Shukla et al., 2000) and other current suggested approaches for WET testing (Bailer et al., 2000).

7.2.3 Phase 2 Analysis–Comparing WET and Bioassessment Results

Phase 2 involves comparing the results of the biological assessments with WET test results and accounting for possible sources of uncertainty in making final conclusions. Agreement/ comparability of results is indicated when WET predicts toxicity and instream biological condition is impaired, or WET predicts no toxicity and instream biological condition is unimpaired. The two types of data can disagree in two ways: WET tests predict instream biological impairment but none is indicated, and WET tests do not predict instream impairment and impairment does exist. Toxicity identification evaluation work or comprehensive stressor identification analysis are probably beyond the scope and resources of the pilot project. However, the supporting information proposed to be collected in this design (e.g., land uses, frequency of plant upsets, upstream condition relative to reference) should enable some clarification of the cause of disagreements. A more structured stressor identification process may be appropriate for some facilities.

The central analysis for this phase is polytomous regression of downstream state (3 states: degradation, no change, improvement) against frequency, magnitude, and type of discharge. A polytomous regression is analogous to the logistic regression used in Diamond et al.(1999), but with more than 2 categories (states in the model). Logistic regression models the frequency or probability of state (degraded, undegraded) as predicted by the explanatory variables (Montgomery and Peck, 1992). Explanatory variables include not only discharge information, but also covariates such as habitat condition, stream size, and upstream water quality.

If there are substantial numbers of downstream sites that are actually improved over the upstream sites, data analysis also should include log linear models (a contingency table analysis) and, if the data can support it, polytomous regression models (multiple-category analog to logistic regression).

For each biological assemblage (i.e., macroinvertebrate, periphyton, and fish), there should be an analysis with each WET endpoint to determine concordance of results. If three alternates are evaluated for each type of WET endpoint, there may be as many as nine comparisons for each WET endpoint (3 endpoint alternatives × 3 types of assessments). With at least seven unique types of WET endpoints in the study design (i.e., fish acute, daphnid acute, fish chronic survival, fish growth, daphnid chronic survival, daphnid reproduction, and algal growth), this could mean 9 × 7 or 56 comparisons for analysis.

With 56 comparisons, spurious results would be expected in 2 or 3 cases just by chance at a $p = 0.05$ level. To minimize purely statistical coincidences, therefore, two additional steps should be taken: (1) compare all bioassessment results with each other to determine a weight of evidence for impairment or non-impairment; then compare all WET endpoints to the overall assessments for that site; and (2) compare each WET endpoint to its corresponding biological assessment: algal growth with periphyton assessment, *Ceriodaphnia* with macroinvertebrates, and *Pimephales* with fish assessment. Both steps will reduce the number of comparisons in an unbiased way and supply additional interpretive information.

Table 7-1. Estimated standard deviations of benthic macroinvertebrate metrics from Florida streams (Source: Barbour et al., 1996b).

Metric	S.D.[a]	90% C.I.[b]
Total taxa	8.90	14.7
Ephemeroptera/Plecoptera/ Trichoptera (EPT) Taxa	7.82	12.9
Florida Index	12.21	20.1
Percent of dominant taxon	6.68	11.0
Percent Diptera	7.70	12.7
Percent filter-gatherers	17.91	29.6
Chironomidae genera	13.11	21.6
Modified Stream Condition Index (SCI) (mean of metric scores)	5.87	9.68

[a] N=18 sites with repeated samples, standard deviation (S.D.) estimated from mean squared error of analysis of variance. All metrics and index scored on 100-point scale, so S.D. estimates also are expressed as percentages.

[b] The 90% CI was calculated with the assumption that the estimated S.D. holds for large sample sizes.

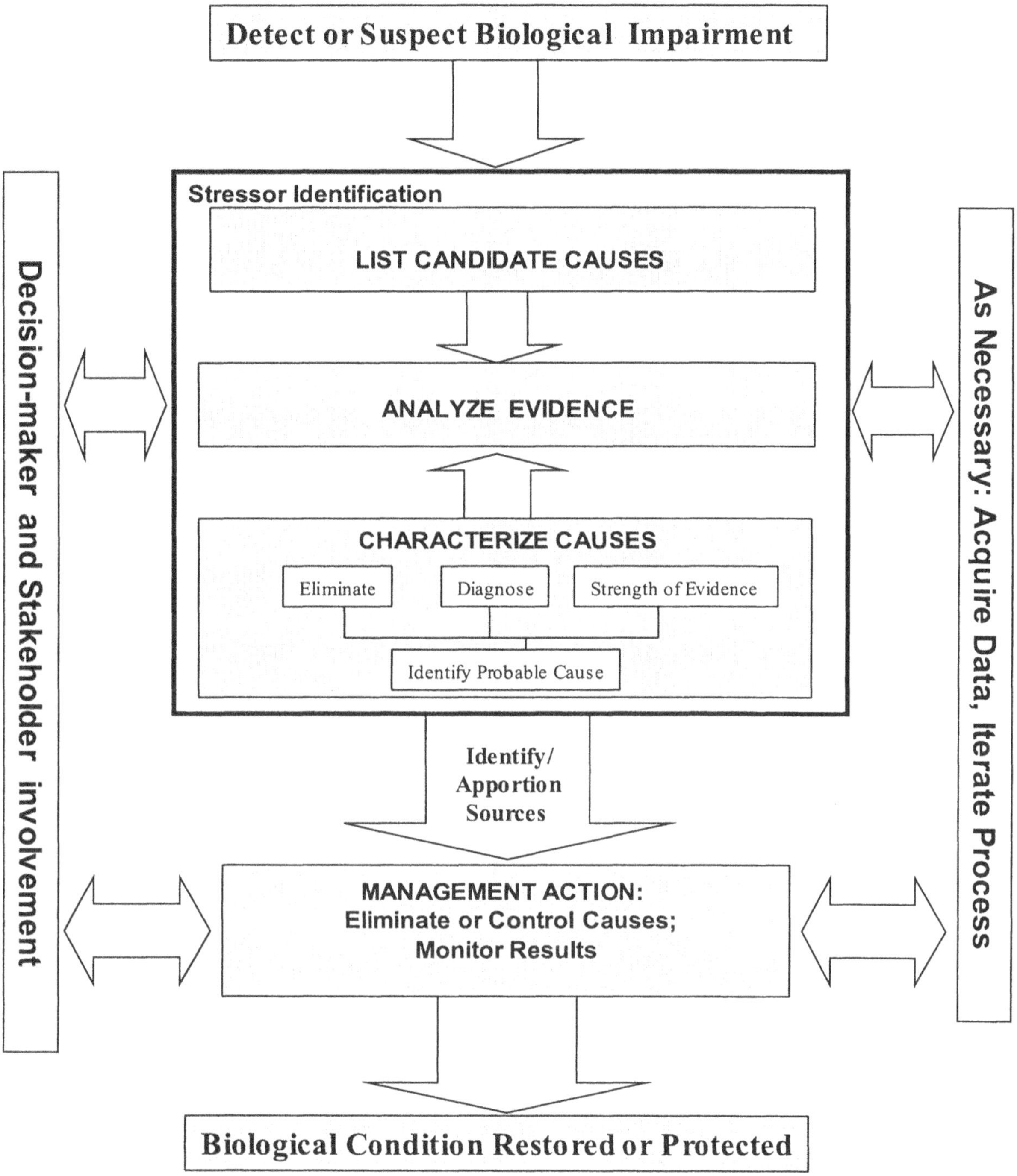

Figure 7-1. The Management Context of the Stressor Identification Process (Source: EPA, 2000d)

APPENDIX A

WET TESTING

APPENDIX A: WET TESTING

Chronic WET testing using *Ceriodaphnia dubia*

Less than 24-hour daphnids are exposed to a minimum of five effluent concentrations and a control. Organisms are exposed to the test conditions until 60% of the surviving control organisms have three broods of offspring for a maximum of eight days. Table A-1 summarizes the test conditions and acceptability criteria for *C. dubia*. The test provides dose response information on the effluent in terms of the effects it has on the survival and reproduction of the organism. The results are expressed as both the highest effluent concentration in which there is no significant effect compared to the control (NOEC) and the concentration at which a 25% decrease in survival or reproduction occurs (IC_{25}). Complete testing methods and procedures are given in USEPA, 2002.

Chronic WET testing using *Pimephales promelas* (fathead minnow)

Newly hatched fathead minnow larvae (> 24hr old) are exposed to a minimum of five effluent concentrations. The test exposure time is seven days. Table A-2 summarizes the test conditions and acceptability criteria for *P. promelas*. The test provides dose response information on the effluent concentration that effects survival and growth of the organism. The results are expressed as the highest effluent concentration in which there is no significant effect compared to the control (NOEC) and the concentration at which a 25% decrease in survival or biomass occurs (IC_{25}). Complete testing methods and procedures are given in USEPA, 2002.

Chronic testing using *Selanastrum capricornutum* (green alga)

A *Selanastrum capricornutum* population is exposed to a minimum of five effluent concentrations and a control in a static system for 96h. Table A-3 summarizes the test conditions and acceptability criteria for *S. capricornutum*. The test provides dose-response information on the effect the effluent has on the growth and development of the population by measuring the changes in cell density biomass, chlorophyll content, or absorbance. Complete testing methods and procedures are given in USEPA, 2002.

Table A-1. Summary Of Test Conditions And Test Acceptability Criteria For Daphnid, *Ceriodaphnia dubia*, Survival And Reproduction Toxicity Tests With Effluents And Receiving Waters (Source: USEPA, 2002)

1. Test type:	Static renewal
2. Temperature (°C):	25 ± 1°C
3. Light quality:	Ambient laboratory illumination
4. Light intensity:	10-20 $\mu E/m^2/s$, or 50-100 ft-c (ambient laboratory levels)
5. Photoperiod:	16 h light, 8 h dark
6. Test chamber size:	30 mL (minimum)
7. Test solution volume:	15 mL (minimum)
8. Renewal of test solutions:	Daily
9. Age of test organisms:	Less than 24 h; and all released within a 8-h period
10. No. neonates per test chamber:	1
11. No. replicate test chambers per concentration:	10
12. No. neonates per test concentration:	10
13. Feeding regime:	Feed 0.1 mL each of YCT and algal suspension per test chamber daily
14. Cleaning:	Use freshly cleaned glass beakers or new plastic cups daily
15. Aeration:	None
16. Dilution water:	Uncontaminated source of receiving or other natural water, synthetic water prepared using MILLIPORE MILLI-Q® or equivalent deionized water and reagent grade chemicals or DMW (see Section 7, Dilution Water)
17. Test concentrations:	Effluents: Minimum of 5 and a control Receiving Water: 100% receiving water or minimum of 5 and a control
18. Dilution factor:	Effluents: ≥ 0.5 Receiving Waters: None or ≥ 0.5

Table A-1 (continued). Summary Of Test Conditions And Test Acceptability Criteria For Daphnid, *Ceriodaphnia dubia*, Survival And Reproduction Toxicity Tests With Effluents And Receiving Waters (Source: USEPA, 2002)

19. Test duration:	Until 60% of surviving control organisms have three broods (maximum test duration 8 days)
20. Endpoints:	Survival and reproduction
21. Test acceptability criteria:	80% or greater survival and an average of 15 or more young per surviving female in the control solutions. 60% of surviving control organisms must produce three broods.
22. Sampling requirements:	For on-site tests, samples collected daily, and used within 24 h of the time they are removed from the sampling device. For off-site tests, a minimum of three samples collected on days one, three, and five with a maximum holding time of 36 h before first use (see Section 8, Effluent and Receiving Water Sampling, Sample Handling, and Sample Preparation for Toxicity Tests, Subsection 8.5.4)
23. Sample volume required:	1 L/day

Table A-2. Summary Of Test Conditions And Test Acceptability Criteria For Fathead Minnow, *Pimephales promelas*, Larval Survival And Growth Toxicity Tests With Effluents And Receiving Waters (Source: USEPA, 2002)

1. Test type:	Static renewal
2. Temperature (°C):	25 ± 1° C
3. Light quality:	Ambient laboratory illumination
4. Light intensity:	10-20 μE/m^2/s (50-100 ft-c)(ambient laboratory levels)
5. Photoperiod:	16 h light, 8 h darkness
6. Test chamber size:	500 mL (minimum)
7. Test solution volume:	250 mL (minimum)
8. Renewal of test solutions:	Daily
9. Age of test organisms:	Newly hatched larvae less than 24 h old. If shipped, not more than 48 h old, 24 h range in age
10. No. larvae per test chamber:	15 (minimum of 10)
11. No. replicate chambers per concentration:	4 (minimum of 3)
12. No. larvae per concentration:	60 (minimum of 30)
13. Source of food:	Newly hatched *Artemia* nauplii (less than 24 h old)
14. Feeding regime:	Feed 0.1 g newly hatched (less than 24-h old) brine shrimp nauplii three times daily at 4-h intervals or, as a minimum, 0.15 g twice daily, 6 h between feedings (at the beginning of the work day prior to renewal, and at the end of the work day following renewal). Sufficient nauplii are added to provide an excess. Larvae fish are not fed during the final 12 h of the test

Table A-2 (continued). Summary Of Test Conditions And Test Acceptability Criteria For Fathead Minnow, *Pimephales promelas*, Larval Survival And Growth Toxicity Tests With Effluents And Receiving Waters (Source: USEPA, 2002)

15. Cleaning:	Siphon daily, immediately before test solution renewal
16. Aeration:	None, unless DO concentration falls below 4.0 mg/L. Rate should not exceed 100 bubbles/min
17. Dilution water:	Uncontaminated source of receiving or other natural water, synthetic water prepared using MILLIPORE MILLI-Q or equivalent deionized water and reagent grade chemicals, or DMW (see Section 7, Dilution Water)
18. Test concentrations:	Effluents: Minimum of 5 and a control Receiving Water: 100% receiving water or minimum of 5 and a control
19. Dilution factor:	Effluents: > 0.5 Receiving waters: None or ≥ 0.5
20. Test duration:	7 days
21. Endpoints:	Survival and growth (weight)
22. Test acceptability criteria:	80% or greater survival in controls; average dry weight per surviving organism in control chambers equals or exceeds 0.25 mg
23. Sampling requirements:	For on-site tests, samples collected daily, and used within 24 h of the time they are removed from the sampling device; For off-site tests, a minimum of three samples collected on days one, three and five with a maximum holding time of 36 h before first use (see Section 8)
24. Sample volume required:	2.5 L/day

Table A-3. Summary Of Test Conditions And Test Acceptability Criteria For Green Alga, *Selenastrum capricornutum*, Growth Toxicity Tests With Effluents And Receiving Waters (Source: USEPA, 2002)

1. Test type:	Static non-renewal
2. Temperature:	25 ± 1°C
3. Light quality:	"Cool white" fluorescent lighting
4. Light intensity:	86 ± 8.6 μE/m^2/s (400 ± 40 ft-c or 4306 lux)
5. Photoperiod:	Continuous illumination
6. Test chamber size:	125 mL or 250 mL
7. Test solution volume:	50 mL or 100 mL [1]
8. Renewal of test solutions:	None
9. Age of test organisms:	4 to 7 days
10. Initial cell density in test chambers:	10,000 cells/mL
11. No. replicate chambers per concentration:	4 (minimum of 3)
12. Shaking rate:	100 cpm continuous, or twice daily by hand
13. Aeration:	None
14. Dilution water:	Algal stock culture medium, enriched uncontaminated source of receiving or other natural water, synthetic water prepared using MILLIPORE MILLI-Q® or equivalent deionized water and reagent grade chemicals, or DMW (see Section 7, Dilution Water)

[1] For tests not continuously shaken use 25 mL in 125 mL flasks and 50 mL in 250 mL flasks.

Table A-3 (continued). **Summary Of Test Conditions And Test Acceptability Criteria For Green Alga, *Selenastrum capricornutum*, Growth Toxicity Tests With Effluents And Receiving Waters (Source: USEPA, 2002)**

15. Test concentrations:	Effluents: Minimum of 5 and a control Receiving Water: 100% receiving water or minimum of 5 and a control
16. Test dilution factor:	Effluents: ≥ 0.5 Receiving Waters: None or ≥ 0.5
17. Test duration:	96 h
18. Endpoint:	Growth (cell counts, chlorophyll fluorescence, absorbance, biomass)
19. Test acceptability criteria:	1 X 10^6 cells/mL with EDTA or 2 X 10^5 cells/mL without EDTA in the controls: Variability of controls should not exceed 20%
20. Sampling requirements:	For on-site tests, one sample collected at test initiation, and used within 24 h of the time it is removed from the sampling device. For off-site tests, holding time must not exceed 36 h (see Section 8, Effluent and Receiving Water Sampling, Sample Handling, and Sample Preparation for Toxicity Tests, Subsection 8.5.4)
21. Sample volume required:	1 or 2 L depending on test volume

APPENDIX B

SUMMARY OF BIOASSESSMENT PROTOCOLS

APPENDIX B: SUMMARY OF BIOASSESSMENT PROTOCOLS

Bioassessment for Macroinvertebrates, Fish, and Periphyton

Field sampling of macroinvertebrates, fish, and periphyton should follow procedures used in USEPA' Rapid Bioassessment Protocols for use in Streams and Wadeable Rivers (Barbour et al., 1999). Table B-1 provides an overview of field sampling procedures for standard (multihabitat) bioassessment of macroinvertebrates. Table B-2 provides an overview of sampling procedures for artificial substrates (Hester-Dendy). In following the study design for the pilot, each facility will perform bioassessments for macroinvertebrates using both the standard multihabitat bioassessment and artificial substrates (Hester Dendy). Table B-3 provides an overview of sampling procedures for fish. Table B-4 provides an overview of sampling procedures for periphyton. Procedures provided in this appendix are not meant to be used as Standard Operating Procedures but to give a brief summary of each sampling technique. Standard Operation Procedures for field collection, lab processing, and data analysis for each assemblage will be provided in the Quality Assurance Project Plan at the initiation of the study.

Table B-1. Field Sampling Procedures for Macroinvertebrates – Standard Bioassessment (Adapted from Barbour et al., 1999)

1. A 100 m reach that is representative of the characteristics of the stream should be selected. Whenever possible, the area should be at least 100 m upstream from any road or bridge crossing to minimize its effect on stream velocity, depth and overall habitat quality. There should be no major tributaries discharging to the stream in the study area.

2. Before sampling, complete the physical/chemical field sheet (see Chapter 5; Appendix A-1, Form 1 of Barbour et al., 1999) to document site description, weather conditions, and land use. After sampling, review this information for accuracy and completeness.

3. Draw a map of the sampling reach. This map should include in-stream attributes (e.g., riffles, falls, fallen trees, pools, bends, etc.) and important structures, plants, and attributes of the bank and near stream areas. Use an arrow to indicate the direction of flow. Indicate the areas that were sampled for macroinvertebrates on the map. Approximate "river mile" to sampling reach for probable use in data management of the water resource agency. If available, use hand-held GPS for latitude and longitude determination taken at the furthest downstream point of the sampling reach.

4. Different types of habitat are to be sampled in approximate proportion to their representation of surface area of the total macroinvertebrate habitat in the reach. For example, if snags comprise 50% of the habitat in a reach and riffles comprise 20%, then 10 jabs should be taken in snag material and 4 jabs should be take in riffle areas. The remainder of the jabs (6) would be taken in any remaining habitat type. Habitat types contributing less than 5% of the stable habitat in the stream reach should not be sampled. In this case, allocate the remaining jabs proportionately among the predominant substrates. The number of jabs taken in each habitat type should be recorded on the field data sheet.

5. Sampling begins at the downstream end of the reach and proceeds upstream. A total of 20 jabs or kicks will be taken over the length of the reach; a single *jab* consists of forcefully thrusting the net into a productive habitat for a linear distance of 0.5 m. A *kick* is a stationary sampling accomplished by positioning the net and disturbing the substrate for a distance of 0.5 m upstream of the net.

6. The jabs or kicks collected from the multiple habitats will be composited to obtain a single homogeneous sample. Every 3 jabs, more often if necessary, wash the collected material by running clean stream water through the net two to three times. If clogging does occur that may hinder obtaining an appropriate sample, discard the material in the net and redo that portion of the sample in the same habitat type but in a different location. Remove large debris after rinsing and inspecting it for organisms; place any organisms found into the sample container. Do not spend time inspecting small debris in the field.

7. Transfer the sample from the net to sample container(s) and preserve in enough 95% ethanol to cover the sample. Forceps may be needed to remove organisms from the dip net. Place a label indicating the sample identification code or lot number, date, stream name, sampling location, and collector name into the sample container. The outside of the container should include the same information and the words "preservative: 95% ethanol". If more that one container is needed for a sample, each container label should contain all the information for the sample and should be numbered (e.g., 1 of 2, 2 of 2, etc.). This information will be recorded in the "Sample Log" at the biological laboratory (Appendix A-3, Form 2 of Barbour et al., 1999).

Table B-1 (continued). Field Sampling Procedures for Macroinvertebrates – Standard Bioassessment (Adapted from Barbour et al., 1999)

8. Complete the top portion of the "Benthic Macroinvertebrate Field Data Sheet" (Appendix A-3, Form 1 of Barbour et al., 1999), which duplicates the "header" information on the physical/chemical field sheet.

9. Record the percentage of each habitat type in the reach. Note the sampling gear used, and comment on conditions of the sampling, e.g., high flows, treacherous rocks, difficult access to stream, or anything that would indicate adverse sampling conditions.

10. Document observations of aquatic flora and fauna. Make qualitative estimates of macroinvertebrate composition and relative abundance as a cursory estimate of ecosystem health and to check adequacy of sampling.

11. Perform habitat assessment (Appendix A-1, Form 3 of Barbour et al., 1999) after sampling has been completed. Having sampled the various microhabitats and walked the reach helps ensure a more accurate assessment. Conduct the habitat assessment with another team member, if possible.

12. Return samples to laboratory and complete log-in forms (Appendix A-3, Form 2 of Barbour et al., 1999).

Table B-2. Field Sampling Procedure For Artificial Substrates (Modified from MDEQ, 2001)

1. Construct substrate samplers by assembling five 3 × 3 × ¼ inch Masonite plates on a four inch eye bolt with two ¼ inch spaces and two ½ spaces between the plates. The total surface area (exclusive of edges and spacers) for each sampler is approximately 0.05 m^2.
2. To deploy the samplers, suspend using steel aircraft cable attached to any available substrate. If there is not an available substrate to attach the cable, the samplers should be suspended between a small Styrofoam float and a brick anchor. The brick or other anchor should also be attached with cording to the nearest stable substrate at, or above, the water level. Samplers should be suspended in the middle of the photic zone. A 1.2m or less depth is recommended. If the water is less than 2.5m deep, the sampler should be suspended at mid-depth.
3. Deploy three samplers at each station utilizing both sides of the river where possible and placed in areas of varying flow regimes. The samplers should also be placed out of public view when possible.
4. Retrieve the samplers after six weeks exposure time. Use a D-frame net to retrieve the samplers. Place the samplers in a 1L container containing 95% ethanol. Check the net and remove any organisms found on the net and add to the sample container.
5. Label the outside of the container with appropriate information (station name, station location, station number, date, and preservative) and place an internal saple label with identical information into each sample container before securely closing the lid. Rotate the container to ensure that the entire sample is preserved. Use a permanent marker or pencil only on the internal label.
6. Transport samples to laboratory.

Table B-3. Field Sampling Procedures for Fish (Adapted from Barbour et al., 1999)

1. A representative stream reach (see Alternatives for Stream Reach Designation, next page) is selected and measured such that primary physical habitat characteristics of the stream are included within the reach (e.g., riffle, run and pool habitats, when available). The sample reach should be located away from the influences of major tributaries and bridge/road crossings (e.g., sufficiently upstream to decrease influences on overall habitat quality). The exact location (i.e., latitude and longitude) of the downstream limit of the reach must be recorded on each field data sheet. (If a Global Positioning System unit is used to provide location information, the accuracy or design confidence of the unit should be noted.) A habitat assessment and physical/ chemical characterization of water quality should be performed within the same sampling reach (see Chapter 5: Habitat Assessment and Physicochemical Characterization).

2. Collection via electrofishing begins at a shallow riffle, or other physical barrier at the downstream limit of the sample reach, and terminates at a similar barrier at the upstream end of the reach. In the absence of physical barriers, block nets should be set at the upstream and downstream ends of the reach prior to the initiation of any sampling activities.

3. Fish collection procedures commence at the downstream barrier. A minimum 2-person fisheries crew proceeds to electrofish in an upstream direction using a side-to-side or bank-to-bank sweeping technique to maximize area coverage. All wadeable habitats within the reach are sampled via a single pass, which terminates at the upstream barrier. Fish are held in livewells (or buckets) for subsequent identification and enumeration.

4. Sampling efficiency is dependent, at least in part, on water clarity and the field team's ability to see and net the stunned fish. Therefore, each team member should wear polarized sunglasses, and sampling is conducted only during periods of optimal water clarity and flow.

5. All fish (greater than 20 millimeters total length) collected within the sample reach must be identified to species (or subspecies). Specimens that cannot be identified with certainty in the field are preserved in a 10% formalin solution and stored in labeled jars for subsequent laboratory identification (see Section 8.2). A representative voucher collection must be retained for unidentified specimens, very small specimens, new locality records, and/or a particular region. In addition to the unidentified specimen jar, a voucher collection of a subsample of each species identified in the field should be preserved and labeled for subsequent laboratory verification, if necessary. Obviously, species of special concern (e.g., threatened, endangered) should be noted and released *immediately* on site. Labels should contain (at a minimum) location data (verbal description and coordinates), date, collectors' names, and sample identification code and/or station numbers for the particular sampling site. Young-of-the-year fish less than 20 millimeters (total length) are not identified or included in the sample, and are released on site. Specimens that can be identified in the field are counted, examined for external anomalies (i.e., deformities, eroded fins, lesions, and tumors), and recorded on field data sheets. An example of a "Fish Sampling Field Data Sheet" is provided in Appendix A-4, Form 1 of Barbour et al., 1999. Space is available for optional fish length and weight measurements, should a particular program/study require length frequency or biomass data. However, these data *are not required* for the standard
multimetric assessment. Space is allotted on the field data sheets for the *optional* inclusion of measurements (nearest millimeter total length) and weights (nearest gram) for a subsample (to a

maximum 25 specimens) of each species. Although fish length and weight measurements are optional, recording a range of lengths for species encountered may be a useful routine measure. Following the data recording phase of the procedure, specimens that have been identified and processed in the field are released on site to minimize mortality.

6. The data collection phase includes the completion of the top portion of the "Fish Sampling Field Data Sheet" (Appendix A-4, Form 1 of Barbour et al., 1999), which duplicates selected information from the physical/chemical field sheet. Information regarding the sample collection procedures must also be recorded. This includes method of fish capture, start time, ending time, duration of sampling, maximum and mean stream widths. The percentage of each habitat type in the reach is estimated and documented on the data sheet. Comments should include sampling conditions, e.g., visibility, flow, difficult access to stream, or anything that may prove to be valuable information to consider for future sampling events or by personnel unfamiliar with the site.

Table B-4. Field Sampling Procedures for Periphyton

The following procedures for multihabitat sampling of algae have been adapted from the Kentucky and Montana protocols (Kentucky DEP 1993, Bahls 1993). These procedures are recommended when subsequent laboratory assessments of species composition of algal assemblages will be performed.

1. Establish the reach for multihabitat sampling as per the macroinvertebrate protocols. In most cases, the reach required for periphyton sampling will be the same size as the reach required for macroinvertebrate or fish sampling (30-40 stream widths) so that as many algal habitats can be sampled as is practical.

2. Before sampling, complete the physical/chemical field sheet (see Chapter 5; Appendix A-1, Form 1) and the periphyton field data sheet (Appendix A-2, Form 1). Visual estimates or quantitative transect-based assessments can be used to determine the percent coverage of each substrate type and the estimated relative abundance of macrophytes, macroscopic filamentous algae, diatoms and other microscopic algal accumulations (periphyton), and other biota (see section 6.2).

3. Collect algae from all available substrates and habitats. The objective is to collect a single composite sample that is representative of the periphyton assemblage present in the reach. Sample all substrates (Table 6-1) and habitats (riffles, runs, shallow pools, nearshore areas) roughly in proportion to their areal coverage in the reach. Within a stream reach, light, depth, substrate, and current velocity can affect species composition of periphyton assemblages. Changes in species composition of algae among habitats are often evident as changes in color and texture of the periphyton. Small amounts (about 5 mL or less) of subsample from each habitat are usually sufficient. Pick specimens of macroalgae by hand in proportion to their relative abundance in the reach. Combine all samples into a common container.

4. Place all samples into a single water-tight, unbreakable, wide-mouth container. A composite sample measuring four ounces (ca. 125 ml) is sufficient (Bahls, 1993). Add recommended amount of Lugol's (IKI) solution, "M3" fixative, buffered 4% formalin, 2% glutaraldehyde, or other preservative (APHA 1995).

5. Place a permanent label on the outside of the sample container with the following information: waterbody name, location, station number, date, name of collector, and type of preservative. Record this information and relevant ecological information in a field notebook or on the periphyton field data sheet (Appendix A-2, Form 1). Place another label with the same information inside the sample container. (Caution! Lugol's solution and other iodine-based preservatives will turn paper labels black.)

6. After sampling, review the recorded information on all labels and forms for accuracy and completeness.

7. Examine all brushing and scraping tools for residues. Rub them clean and rinse them in distilled water before sampling the next site and before putting them away.

8. Transport samples back to the laboratory in a cooler with ice (keep them cold and dark) and store preserved samples in the dark until they are processed. Be sure to stow samples in a way so that

transport and shifting does not allow samples to leak. When preserved, check preservative every few weeks and replenish as necessary until taxonomic evaluation is completed.

9. Log in all incoming samples (Appendix A-2, Form 2). At a minimum, record sample identification code, date, stream name, sampling location, collector's name, sampling method, and area sampled (if it was determined).

REFERENCES

ASTM. 2000. E729-96 Standard guide for conducting acute toxicity tests with fishes, macroinvertebrates, and amphipods. pp. 218-238. In: *1998 Annual Book of ASTM Standards*, Vol. 11.05, Philadelphia, PA.

Bailer, A.J., M.R. Hughes, D.L. Denton, and J.T. Oris. 2000. An Empirical Comparison of Effective Concentration Estimators for Evaluating Aquatic Toxicity Tests Responses. *Environ. Toxicol. Chem.* 19:141-150.

Baird, R., R. Berger, and J. Gully. 1996. Improvements in point estimation methods and application to controlling aquatic toxicity test reliability. In: D. Grothe, K. Dickson, and D. Reed-Judkins (editors). *Whole Effluent Toxicity Testing: An evaluation of methods and prediction of receiving system impacts*. SETAC Press, Pensacola, FL.

Barbour, M.T., J. Gerritsen, B.D. Snyder, and J.B. Stribling. 1999. *Rapid Bioassessment Protocols for Use in Streams and Wadeable Rivers: Periphyton, Benthic Macroinvertebrates and Fish, Second Edition*. EPA 841-B-99-002. U.S. Environmental Protection Agency; Office of Water; Washington, D.C.

Barbour, M.T., J.M. Diamond, C.O. Yoder. 1996a. Biological assessment strategies: Applications and Limitations. Pages 245-270, In: D.R. Grothe, K.L. Dickson, and D.K. Reed-Judkins (editors). *Whole effluent toxicity testing: An evaluation of methods and prediction of receiving system impacts*, SETAC Press, Pensacola, FL.

Barbour, M.T., J. Gerritsen, and J.S. White. 1996b. Development of a Stream Condition Index for Florida. Prepared for the Florida Department of Environmental Protection, Tallahassee, FL.

Barbour M.T., J.R. Stribling, and J.R. Karr. 1995. Multimetric approach for establishing biocriteria and measuring biological condition. Pp. 63-77. In: Davis W.S. and T.P. Simon (editors). *Biological assessment and criteria: tools for water resource planning and decision making*. CRC Press, Boca Raton, FL.

Birge, W., J. Black, T. Short, and A. Westerman. 1989. A comparative ecological and toxicological investigation of a secondary wastewater treatment plant effluent and its receiving stream. *Environ. Toxicol. Chem.* 8:437-450.

Bode, R.W. and M.A. Novak. 1995. Development and application of biological impairment criteria for rivers and streams in New York State. Pages 97-107, In: W.S. Davis and T.P. Simon (editors). *Biological Assessment and Criteria: Tools for Water Resource Planning and Decision Making*. Lewis, Boca Raton, FL.

Cairns, J., Jr. 1986. What is meant by validation of predictions based on laboratory toxicity tests? *Hydrobiologia* 137:271-278.

Diamond, J. M., M. T. Barbour, and J. B. Stribling. 1996. Characterizing and comparing bioassessment methods and their results: a perspective. *Journal of the North American Benthological Society* 15:713-727.

Diamond, J. and C. Daley. 2000. What is the relationship between whole effluent toxicity result and biological condition? *Environ Toxicol Chem* 19:158-168.

Diamond, J., C. Daley, T. Moore. 1999. Evaluating Whole Effluent Toxicity Testing as an Indicator of Instream Biological Conditions. Water Environment Research Foundation Report, 95-HHE-1, Alexandria, VA.

Dickson K., W. Waller, J. Kennedy, L. Ammann. 1992. Assessing the relationship between ambient toxicity and instream biological response. *Environ. Toxicol. Chem.* 11:1307-1322.

Dyer, Scott D. and X. Wang. 2002: A Comparison Of Stream Biological Responses To Discharge From Wastewater Treatment Plants In High And Low Population Density Areas. *Environ. Toxico. Chem.* 21:1065–1075.

Eagelson, K., D. Lenat, L. Ausley, F. Winborne. 1990. Comparison of measured instream biological responses with responses predicated by *Ceriodaphnia* chronic toxicity tests. *Environ. Toxicol. Chem.* 9:1019-1028.

Faith, D.P., C.L. Humphrey, and P.L. Dostine. 1991. Statistical Power and BACI Designs in Biological Monitoring: Comparative Evaluation of Measures of Community Dissimilarity Based on Benthic Macroinvertebrate Communities in Rockhole Mine Creek, Northern Territory, Australia. *Aus. J. Mar. Freshwater Res.* 42:589-602.

Gibson G.R., M.T. Barbour, J.B. Stribling, J. Gerritsen, J.R. Karr. 1996. Biological criteria technical guidance for streams and small rivers. EPA 822-B-96-001. US Environmental Protection Agency.

Great Lakes National Program Office (GLNPO). 1994. Assessment and remediation of Contaminated Sediments (ARCS) Program. Assessment Guidance Document. EPA 905-B94-002, Great Lakes National Program Office, Chicago, IL

Grothe, D.R., K.L. Dickson, and D.K. Reed-Judkins (editors). 1996. *Whole effluent toxicity testing: An evaluation of methods and prediction of receiving system impacts*, SETAC Press, Pensacola, FL

Intergovernmental Task Force on Monitoring (ITFM). 1995. The Strategy for Improving Water Quality Monitoring in the United States. Technical Appendices and Final Report. Intergovernmental Task Force on Monitoring Water Quality. Office of Water Data Coordinator, U.S. Geological Survey, Reston, VA. Open-File Report 95-742.

Karr, J. and E. Chu. 1999. *Restoring life in running waters*. Island Press, Washington, D.C.

Keith, L.H. 1991. *Environmental Sampling and Analysis. A Practical Guide*. Lewis Publishers Inc., Chelsea, MI.

Klemm, D.J., P.A. Lewis, F. Fulk, and J.M. Lazorchak. 1990. Macroinvertebrate Field and Laboratory Methods for Evaluating the Biological Integrity of Surface Waters. US Environmental Protection Agency, EPA-600-4-90-030, Environmental Monitoring and Support Laboratory, Cincinnati, OH

La Point, T.W. and W.T. Waller. 2000. Field assessments in conjunction with whole effluent toxicity testing. *Environ Toxicol Chem* 19:14-24.

Mackie, G.L. 2001. *Applied aquatic ecosystem concepts*. Kendall/Hunt Publishing Company. Dubuque, IA.

Marcus, M. and L. McDonald. 1992. Evaluating the statistical bases for relating receiving water impacts to effluent and ambient toxicities. *Environmental Toxicolology and Chemistry* 11:1389-1402.

Markle, P.J., J.R. Gully, R.B. Baird, K.M. Nakada, and J.P. Bottomley. 2000. Effects of several variables on whole effluent toxicity test performance and interpretation. *Env. Toxicol. Chem.* 19(1):123-132.

Mississippi Department of Environmental Quality (MDEQ). 2001. Quality Assurance Project Plan for 303(d) for Development of Biological Indicators for Streams and Wadeable Rivers in the Mississippi Delta. Pilot Study. December 5, 2001. Mississippi Department of Environmental Quality. Jackson, MS. (For further information, contact Randy Reed [601-961-5158]).

Mississippi Department of Environmental Quality (MDEQ). 2002. Development and Application of the Mississippi Benthic Index of Stream Quality (M-BISQ). Draft, October 4, 2002. Prepared by Tetra Tech, Inc., Owings Mills, MD, for the Mississippi Department of Environmental Quality, Office of Pollution Control, Jackson, MS. (For further information, contact Randy Reed [601-961-5158]).

Mississippi Department of Environmental Quality (MDEQ). 2003. Quality Assurance Project Plan for 303 (d) List Assessment and Calibration of the Index of Biological Integrity. Phase 3. Mississippi Department of Environmental Quality. Jackson, MS. (For further information, contact Randy Reed [601-961-5158]).

Montgomery, D.C. and E.A. Peck. 1992. *Introduction to Linear Regression Analysis. 2d Edition*. John Wiley & Sons, New York.

Moore, T., S. Canton, and M Grimes. 2000. Investigating the incidence of Type 1 errors for chronic whole effluent toxicity testing using *Ceriodaphnia dubia. Environmental Toxicology and Chemistry* 19: 118-122.

Ohio EPA. 1992. Ohio Water Resource Inventory. Volume 1, Status and Trends. E. T. Rankin, C.O. Yoder, and D.A., Mishne (editors). Ohio Environmental Protection Agency, Division of Water Quality Planning and Assessment, Columbus, OH.

Ohio EPA. 1990. Ohio Water Resource Inventory. Volume 1, Status and Trends. E. T. Rankin, C.O. Yoder, and D.A., Mishne (editors). Ohio Environmental Protection Agency, Division of Water Quality Planning and Assessment, Columbus, OH.

Shukla, R., Q. Wang, F. Fulk, C. Deng, and D. Denton, 2000. Bioequivalence approach for whole effluent toxicity testing. *Environmental Toxicology and Chemistry* 19:169-174.

Smith, E.P., D.R. Orvos, and J. Cairns, Jr. 1993. Impact assessment using the before-after-control-impact (BACI) model: concerns and comments. *Can. J. Fish. Aquat. Sci.* 50:627-637.

Snyder, B.D., J.B. Stribling, M.T. Barbour, and C.L. Missimer. 1998. Integrating Assessments of Fish and Macroinvertebrate Assemblages and Physical Habitat Condition. Chapter 26, In: T.P. Simon (editor). *Assessing the Sustainability and Biological Integrity of Water Resource Quality Using Fish Assemblages.* CRC Press, Boca Raton, FL.

Southerland, M. and J.B. Stribling. 1995. The Status of Biocriteria Program Development in States. Chapter 7, In: Davis, W. and T. Simon (editors). *Biological Assessment and Criteria: Tools for Water Resource Planning and Decisionmaking.* Lewis Publishers. Boca Raton, FL.

Sprague, J.B. 1995. Review of methods for sublethal aquatic toxicity tests relevant to the Canadian metal-mining industry, and design of field validation programs. Aquatic Effects Technology Evaluation Program, Canadian Centre for Mineral and Energy Technology, Department of Natural Resources Canada, Ottawa.

Stribling, J. B., B. Jessup, and D. Bressler. (*in review*). Performance characteristics of biological methods: the precision of field sampling and taxonomic identifications. Submitted to the Journal of the North American Benthological Society.

Taylor, J. K. 1988. Defining the Accuracy, Precision, and Confidence Limits of Sample Data. Chapter 6, pages 101-107. In: L. H. Keith (editor), *Principles of Environmental Sampling.* ACS Professional Reference Book. American Chemical Society. Washington, DC.

ToxCalc, Version 5.0.23D. Tidepool Scientific Software, McKinleyville, CA.

Underwood, A.J. 1994. On beyond BACI: sampling designs that might reliably detect environmental disturbances. *Ecol. Applic.* 4:3-15.

URS Corporation, CDM Inc. 2002. Habitat Characterization Study Final Report. Arid West Research Project. Phoenix, AZ.

USEPA. 2002. Short-term Methods for Estimating the Chronic Toxicity of Effluents and Receiving Waters to Freshwater Organisms. Fourth Edition. EPA-821-R-02-013. Office of Water, Washington, D.C.

USEPA. 2000a. Guidance for the Data Quality Objectives Process. EPA QA/G-4. EPA/600/R-96/055. Office of Environmental Information, Washington, D.C.

USEPA. 2000b. Method guidance and recommendations for whole effluent toxicity (WET) testing (40 CFR Part 136). EPA-821-B-00-004, Office of Water, Washington, D.C.

USEPA. 2000c. Preliminary report: Interlaboratory variability study of EPA short-term chronic and acute whole effluent toxicity test methods. EPA-821-R-00-028A, Office of Water, Washington, D.C.

USEPA. 2000d. Stressor Identification Guidance Document. EPA/822/B-00/025. Office of Water/Office of Research and Development; Washington, D.C.

USEPA. 1995. Generic quality assurance project plan guidance for programs using community-level biological assessment in streams and rivers. EPA 841-B-95-004. Office of Water, US Environmental Protection Agency, Washington, DC.

USEPA. 1994. Short-term methods for estimating the chronic toxicity of effluents and receiving water to marine and freshwater organisms. Second Edition. EPA-600-4-91-003. Office of Water, Washington, D.C.

USEPA. 1986. Development of data quality objectives. Descriptions of stages I and II. Prepared by the EPA Quality Assurance Management staff. Office of Research and Development, US Environmental Protection Agency, Washington, DC.

USEPA. 1984. The development of data quality objectives. Prepared by the EPA Quality Assurance Management staff, Office of Research and Development, US Environmental Protection Agency, Washington, DC.

Waller, W.T., L.P. Amman, W.J. Birge, K.L. Dickson, P.B. Dorn, N.E. LeBlanc, D.I. Mount, B.R. Parkhurst, R.H. Preston, S.C. Schimmel, A. Spacie, G.B. Thursby.. 1996. Predicting instream effects from WET tests. pp 271-286, In: Grothe, D.R., Dickson K.L., Reed-Judkins, D.K. (editors). *Whole Effluent Toxicity Testing: An Evaluation of Methods and Prediction of Receiving System Impacts.* Society of Environmental Toxicology and Chemistry, Pensacola, FL..

Warren-Hicks, W. J., and B. R. Parkhurst. 1996. Issues in whole effluent toxicity test uncertainty analyses. Discussion-Initiation Paper 6.3. pp 180-189, In: Grothe, D. R., Dickson, K. L., Reed-Judkins, D. K.(editors). *Whole effluent toxicity testing: an evaluation of methods and prediction of receiving system impacts.* SETAC Pellston Workshop on Whole Effluent Toxicity; 1995 Sep 16-25; Pellston, MI. Pensacola, FL; SETAC Press. 340 pp.

Warren-Hicks, W., B. Parkhurst, D. Moore, R. Teed, R. Baird, R. Berger, D. Denton, J. Plett. 2000. Assessment of whole effluent toxicity test variability: partitioning sources of variability. *Environ Tox Chem* 19:94-104.

Yoder, C.O., and E.T. Rankin. 1995. Biological response signatures and the area of degradation value: New tools for interpreting multimetric data. pp.263-286, In: W.S. Davis and T.P. Simon (editors). *Biological Assessment and Criteria: Tools for Water Resource Planning and Decision Making*. Lewis, Boca Raton, FL

WERF SUBSCRIBERS

Wastewater Utility

Alabama
Montgomery Water Works & Sanitary Sewer Board

Alaska
Anchorage Water & Wastewater Utility

Arizona
Gila Resources
Glendale, City of, Utilities Department
Mesa, City of
Phoenix Water Services Department
Pima County Wastewater Management

Arkansas
Little Rock Wastewater Utility

California
Calaveras County Water District
Central Contra Costa Sanitary District
Contra Costa Water District
Crestline Sanitation District
Delta Diablo Sanitation District
Dublin San Ramon Services District
East Bay Municipal Utility District
El Dorado Irrigation District
Escondido, City of
Fairfield-Suisun Sewer District
Irvine Ranch Water District
Las Virgenes Municipal Water District
Lodi, City of
Los Angeles, City of
Los Angeles County, Sanitation Districts of
Napa Sanitation District
Orange County Sanitation District
Palo Alto, City of
Riverside, City of
Sacramento Regional County Sanitation District
San Diego Metropolitan Wastewater Department, City of
San Francisco, City & County of
San Jose, City of
Santa Barbara, City of
Santa Rosa, City of
South Bayside System Authority
South Orange County Wastewater Authority
Union Sanitary District
West Valley Sanitation District

Colorado
Boulder, City of
Colorado Springs, City of
Littleton/Englewood Water Pollution Control Plant
Metro Wastewater Reclamation District, Denver

Connecticut
New Haven, City of, WPCA

District of Columbia
District of Columbia Water & Sewer Authority

Florida
Broward, County of
Fort Lauderdale, City of
Gainesville Regional Utilities
JEA
Kissimmee, City of, Department of Water Resources
Miami-Dade Water & Sewer Authority
Orange County Utilities Department
Orlando, City of
Reedy Creek Improvement District
Seminole County Environmental Services
St. Petersburg, City of
Stuart Public Utilities
Tallahassee, City of
Tampa, City of
West Palm Beach, City of

Georgia
Atlanta Wastewater Services
Augusta, City of
Clayton County Water Authority
Cobb County Water System
Columbus Water Works
Fulton County
Gwinnett County Department of Public Utilities
Macon Water Authority
Savannah, City of

Hawaii
Honolulu, City and County of

Idaho
Pocatello, City of

Illinois
American Bottoms Wastewater Treatment Plant
Dupage, County of
Greater Peoria Sanitary District
Kankakee River Metropolitan Agency
Metropolitan Water Reclamation District of Greater Chicago
Wheaton Sanitary District

Indiana
Fort Wayne, City of
Indianapolis, City of

Iowa
Ames, City of
Des Moines Metro Wastewater Reclamation Authority

Kansas
Johnson County Unified Wastewater Districts
Unified Government of Wyandotte County/Kansas City, City of

Kentucky
Louisville & Jefferson County Metropolitan Sewer District

Louisiana
Sewerage & Water Board of New Orleans

Maine
Bangor, City of
Portland Water District

Maryland
Anne Arundel County Bureau of Utility Operations
Howard County Department of Public Works
Washington Suburban Sanitary Commission

Massachusetts
Boston Water & Sewer Commission
Upper Blackstone Water Pollution Abatement District

Michigan
Ann Arbor, City of
Detroit, City of
Grand Rapids, City of
Holland Board of Public Works
Lansing, City of
Owosso Mid-Shiawassee County WWTP
Saginaw, City of
Wayne County Department of Environment
Wyoming, City of

Missouri
Independence, City of
Kansas City Missouri Water Services Department
Little Blue Valley Sewer District
Metropolitan St. Louis Sewer District

Nebraska
Lincoln Wastewater System

Nevada
Henderson, City of

New Jersey
Passaic Valley Sewerage Commissioners

New York
New York City Department of Environmental Protection

North Carolina
Cary, Town of
Charlotte/Mecklenburg Utilities
Durham, City of
Metropolitan Sewerage District of Buncombe County
Orange Water & Sewer Authority

Ohio
Akron, City of
Butler County Department of Environmental Services
Columbus, City of
Metropolitan Sewer District of Greater Cincinnati
Northeast Ohio Regional Sewer District

Oklahoma
Tulsa, City of

Oregon

Clean Water Services

Eugene/Springfield Water Pollution Control

Water Environment Services

Pennsylvania

Philadelphia, City of

University Area Joint Authority, State College

South Carolina

Charleston Commissioners of Public Works

Mount Pleasant Waterworks & Sewer Commission

Spartanburg Sanitary Sewer District

Tennessee

Cleveland, City of

Knoxville Utilities Board

Murfreesboro Water & Sewer Department

Nashville Metro Water Services

Texas

Amarillo, City of

Austin, City of

Dallas Water Utilities

Denton, City of

El Paso Water Utilities

Fort Worth, City of

Gulf Coast Waste Disposal Authority

Houston, City of

San Antonio Water System

Trinity River Authority

Utah

Salt Lake City Corporation

Virginia

Alexandria Sanitation Authority

Arlington, County of

Fairfax County Virginia

Hampton Roads Sanitation District

Henrico, County of

Hopewell Regional Wastewater Treatment Facility

Loudoun County Sanitation Authority

Lynchburg Regional WWTP

Prince William County Service Authority

Richmond, City of

Washington

Edmonds, City of

Everett, City of

King County Department of Natural Resources

Seattle Public Utilities

Wisconsin

Green Bay Metro Sewerage District

Madison Metropolitan Sewerage District

Milwaukee Metropolitan Sewerage District

Racine, City of

Sheboygan Regional Wastewater Treatment

Wausau Water Works

Australia

Sydney Water Corporation

Water Corporation of Western Australia

Canada

Toronto, City of, Ontario

Mexico

Servicios de Agua y Drenaje de Monterrey, I.P.D.

United Kingdom

Yorkshire Water Services Limited

Stormwater Utility

California

Monterey, City of†

Santa Rosa, City of†

Colorado

Boulder, City of†

Georgia

Griffin, City of†

Kansas

Overland Park, City of†

Kentucky

Louisville & Jefferson County Metropolitan Sewer District†

Maine

Portland Water District†

North Carolina

Cary, Town of†

Charlotte, City of/Mecklenburg County†

Pennsylvania

Philadelphia, City of†

State

Fresno Metropolitan Flood Control District, Calif.†

Urban Drainage & Flood Control District, Colo.†

Corporate

ADS Environmental Services

The ADVENT Group Inc.

Alan Plummer & Associates

Alden Research Laboratory

Alpine Technology Inc.

Aquateam–Norwegian Water Technology Centre A/S

BaySaver Inc.†

BioVir Inc.

Black & Veatch

Boyle Engineering Corporation

BPR CSO

Brown & Caldwell

Burns & McDonnell

CABE Associates Inc.

The Cadmus Group

Camp Dresser & McKee Inc.

Carollo Engineers Inc.

Carpenter Environmental Associates Inc.

CDS Technologies Inc.

Chemtrac Systems Inc.

CH2M HILL

Clancy Environmental Consultants Inc.

Damon S. Williams Associates, LLC

Earth Tech Inc.

Ecolab Water Care Services

EMA Inc.

The Eshelman Company Inc.

Finkbeiner, Pettis, & Strout Inc.

Frontier Geosciences Inc.

ftn Associates Inc.

Gannett Fleming Inc.

GE Betz

Golder Associates Inc.

Greeley and Hansen LLC

The HACH Company

Hazen & Sawyer, P.C.

HDR Engineering Inc.

HNTB Corporation

HydroQual Inc.

Insituform Technologies Inc.

Institute for Environmental Technology & Industry, Korea

Jacobson Helgoth Consultants Inc.

Jason Consultants Inc.

Jordan, Jones, & Goulding Inc.

KCI Technologies Inc.

Kelly & Weaver, P.C.

Kennedy/Jenks Consultants

Komline Sanderson Engineering Corporation

Lawler, Matusky & Skelly Engineers, LLP

Limno-Tech Inc.

Lombardo Associates Inc.

Malcolm Pirnie Inc.

McKim & Creed

MWH

New England Organics

Odor & Corrosion Technology Consultants Inc. (OCTC)

ONDEO Degremont Inc.

PA Government Services Inc.

Parametrix Inc.

Parsons

Post, Buckley, Schuh & Jernigan

The RETEC Group

R.M. Towill Corporation

Ross & Associates Ltd.

Royce Technologies

SAIC Maritime Technical Group

Sear-Brown

Sierra Environmental Services

SYNAGRO

Tetra Tech Inc.

Trojan Technologies Inc.

URS Corporation

USFilter NATC

Vortechnics Inc.†

Wade-Trim Inc.

Weston Solutions Inc.

Woodard & Curran

WRc/D&B, LLC

WWETCO, LLC†

Industry

American Electric Power

BP

ChevronTexaco Energy Research & Technology Company

The Coca-Cola Company

Dow Chemical Company

DuPont Company

Eastman Chemical Company

Eastman Kodak Company

Merck & Company Inc.

ONDEO Services

Procter & Gamble Company

Severn Trent Services Inc.

Shell Global Solutions

Thames Water Plc

United Water Services LLC

† Charter Stormwater Subscribers

Note: List as of 1/03

WERF Product Order Form

As a benefit of joining the Water Environment Research Foundation, subscribers are entitled to receive one complimentary copy of all final reports and other products. Additional copies are available at cost (usually $10). To order your complimentary copy of a report, please write "free" in the unit price column. WERF keeps track of all orders. If the charge differs from what is shown here, we will call to confirm the total before processing.

Name ______________________ Title ______________________

Organization ______________________

Address ______________________

City ______________ State ________ Zip Code ________ Country ________

Phone ______________ Fax ______________ Email ______________

Stock #	Product	Quantity	Unit Price	Total
			Postage & Handling	
			VA Residents Add 4.5% Sales Tax	
			Canadian Residents Add 7% GST	
			TOTAL	

Method of Payment: (All orders must be prepaid.)

- ☐ Check or Money Order Enclosed
- ☐ Visa ☐ Mastercard ☐ American Express

Account No. ______________________ Exp. Date ________

Signature ______________________

Shipping & Handling:

Amount of Order	United States	Canada & Mexico	All Others
Up to but not more than:	Add:	Add:	Add:
$20.00	$5.00*	$8.00	50% of amount
30.00	5.50	8.00	40% of amount
40.00	6.00	8.00	
50.00	6.50	14.00	
60.00	7.00	14.00	
80.00	8.00	14.00	
100.00	10.00	21.00	
150.00	12.50	28.00	
200.00	15.00	35.00	
More than $200.00	Add 20% of order	Add 20% of order	

*minimum amount for all orders

Note: Please make checks payable to the Water Environment Research Foundation.

To Order (Subscribers Only):

Log on to www.werf.org and click on the "Product Catalog."

Phone: (703) 684-2470
Fax: (703) 299-0742.

WERF
Attn: Subscriber Services
601 Wythe Street
Alexandria, VA 22314-1994

To Order (Non-Subscribers):

Non-subscribers may be able to order WERF publications either through WEF (www.wef.org) or IWAP (www.iwapublishing.com). Visit WERF's website at www.werf.org for details.